I0477822

CONTENTS

CULINARY ANTHROPOLOGY

Gastronomy And Culture For Chefs

Dr Bhaskar Bora

A PERSONAL NOTE FROM THE AUTHOR

Though laden with unexpected trials and hardships, my journey has blossomed into a story of profound transformation—a journey that has led me to discover purpose, not in the towering milestones of success, but in the quiet, tender moments of love, care, and presence. What you hold in your hands is not merely a collection of recipes, but a testament to resilience— a narrative stitched together with threads of struggle, acceptance, and, in time, renewal.

There was a time when the story of my life played out with certainty and clarity, like a symphony where each note was perfectly placed. As a Doctor, my days were woven with the pulse of life itself—healing, comforting, offering hope where none had been. I wore my white coat with pride, for it was not just a garment but a symbol of who I was. The work I did, and the lives I touched, gave meaning to my every breath. My identity was fused with my role as a healer as if I had been born to follow that path.

But life, with its intricate unpredictability, had other plans. In a single, unforeseen moment, the world I had so carefully built was undone—first with a spinal cord injury that took away the physical strength I had always known, and then with the looming shadow of cancer, a reminder of how fragile life truly is. The vibrant world of medicine, where I once found purpose and joy, suddenly slipped beyond my reach. What once was filled with meaning became a void, vast and silent, leaving me to ask the questions I never thought I would need to face.

The bustling hospital hallways were exchanged for the quiet solitude of my home, where I was no longer a "Doctor." My hands, once steady with the knowledge of healing, trembled in the face of an unknown future. Who was I without the title, the purpose, and the work that defined me? I stood at the edge of this new reality, uncertain and untethered, wondering what life could offer beyond what I had known.

In the silence of that transition, I discovered something unexpected. What once seemed like an unfamiliar, distant role—being a disabled husband and a disabled father—became the essence of my existence. And within that shift, I found cooking. What began as an effort to nurture my family soon became a source of healing for me. In the rhythm of chopping, stirring, and tasting, I discovered a new purpose. Cooking became a language through which I reconnected with life, a practice that grounded me when everything else felt adrift and connected to the ones I love.

These past few years spent creating nourishing, simple meals, have been my lifeline—a daily practice of care for the people I love and for myself. Exploring different cuisines, experimenting with flavours, reading about ingredients and techniques—all of it became not just a pastime but a pathway to reclaiming my identity. Through cooking, I found a way forward, one meal at a time.

What I share with you now through this book, and those that will follow, are the lessons learned along the way. They are simple, practical, and grounded in love. These recipes and cooking tips are not adorned with glossy images or extravagant flourishes, but they carry with them the essence of resilience, creativity, and joy. I hope that they bring as much warmth and nourishment to your home as they have to mine, and that in their simplicity, you find a way to savour the moments spent with the ones you cherish.

We cannot control what the universe throws at us, but

how we react to those curveballs defines who we are and what we make of our lives.

CHAPTER 1: INTRODUCTION TO CULINARY ANTHROPOLOGY

Introduction To The Field Of Culinary Anthropology, Its Importance, And Relevance For Chefs

Culinary anthropology sits at the fascinating intersection of food, culture, and social identity, examining how food shapes and is shaped by the cultural, historical, and societal landscapes it inhabits. This field seeks to unravel the layers of meaning behind what people eat, how they prepare it, and the social customs surrounding meals. By studying culinary anthropology, chefs can gain a more profound understanding of the food traditions they engage with, enabling them to approach their culinary practices with greater cultural sensitivity, awareness, and purpose. This perspective enriches their craft, transforming each dish into a cultural artifact that tells a story about the people, places, and historical moments behind it.

In culinary anthropology, the act of cooking is not only about nourishment or flavour; it's a deeply social act that serves as a medium for expressing identity, honouring tradition, and fostering community. For chefs, embracing culinary anthropology allows them to go beyond technique and ingredients to consider the cultural meanings embedded within the foods they work with. It enables them to see a dish not merely as a combination of flavours but as a product of history, migration, and social evolution. Through this understanding, chefs can approach their creations as vehicles of cultural exchange, respect, and storytelling.

The significance of culinary anthropology has grown in recent years as food has become a means of cultural diplomacy, identity formation, and cross-cultural dialogue. In an increasingly globalized world, where chefs have access to a wider variety of ingredients and techniques than ever before, there is also a responsibility to represent these influences authentically and thoughtfully. Culinary anthropology provides a foundation for chefs to explore these global traditions responsibly, ensuring that their creations honour and reflect the origins of the foods they represent. It encourages chefs to avoid reducing rich culinary traditions to mere trends or gimmicks, emphasizing the importance of respecting the cultural histories that food embodies.

Understanding culinary anthropology is particularly valuable for chefs navigating today's culinary landscape, where authenticity and cultural awareness are highly valued. With an appreciation of culinary anthropology, chefs are better equipped to avoid cultural appropriation, instead fostering a genuine connection to the foods they prepare. They can draw upon culinary anthropology to create experiences that honour diverse cultural histories, enriching the dining experience for patrons who seek to learn about and appreciate the world's culinary diversity.

Overview of Key Terms and Concepts in Gastronomy and Anthropology

Culinary anthropology introduces several foundational concepts and terms that are essential for chefs to grasp. Gastronomy is one of the core ideas, often described as the art and science of food. It encompasses not only cooking techniques but also the sensory and aesthetic aspects of dining, such as taste, texture, and presentation. Gastronomy lies at the heart of culinary creativity, bridging the technical with the artistic and allowing chefs to craft dishes that captivate the senses. Through gastronomy, chefs can explore how different cooking methods impact the flavour and texture of ingredients, how presentation enhances the dining experience, and how certain ingredient pairings elicit specific emotional responses from diners.

Anthropology, on the other hand, is the study of humans and their cultures, including how they relate to one another and to the environment. When applied to food, anthropology delves into how food traditions develop, evolve, and intersect across different societies. This perspective invites chefs to see food as more than just sustenance; it's a medium through which individuals and communities express their values, beliefs, and identities. Anthropology helps chefs understand that every dish has a story, shaped by a complex web of social, economic, and environmental factors.

One of the essential concepts within culinary anthropology is foodways—the ways in which people acquire, prepare, and consume food. Foodways encompass everything from the agricultural practices that sustain food production to the social rituals of dining. For chefs, foodways offer insights into how specific ingredients and techniques reflect a culture's relationship with the land, climate, and seasonality. For example, Japan's foodways are deeply influenced by its coastal geography, which has led to a culinary tradition that emphasizes seafood, rice, and the art of subtle flavours. Understanding foodways allows chefs to connect more deeply with the ingredients they use and to consider the environmental factors that influence the flavour, texture, and availability of different foods.

Another important term in culinary anthropology is food symbolism, which refers to the meanings that different cultures assign to specific foods. These meanings may be tied to religious beliefs, historical events, or social customs. For instance, rice in many Asian cultures symbolizes prosperity and community, while in the Western context, bread often represents sustenance and hospitality. By understanding these symbols, chefs can appreciate the significance of the foods they work with, using this knowledge to create dishes that resonate with cultural meaning. This sensitivity to food symbolism can also guide chefs in crafting menus that respect the cultural contexts of their dishes, enhancing the authenticity and depth of the dining experience.

A third key concept is culinary syncretism, or the blending of culinary traditions, which often occurs when different cultural groups come into contact through trade, migration, or colonization. Culinary syncretism has given rise to some of the world's most beloved dishes, such as the spicy curries of Southeast Asia, influenced by Indian and Chinese flavours, or the fusion of European, African, and Native American flavours in Creole cuisine. For chefs, culinary syncretism is a valuable reminder of the dynamic and evolving nature of food, encouraging them to view their own culinary creations as part of a broader, ongoing story of cultural exchange and adaptation.

Finally, chefs should be aware of the concept of food sovereignty, which emphasizes the rights of communities to control their food systems and make decisions about what they grow, eat, and share. Food sovereignty is especially relevant in the context of indigenous culinary traditions, where many communities are seeking to preserve their food heritage in the face of globalization. For chefs, an awareness of food sovereignty encourages responsible sourcing, respect for traditional food practices, and support for local food systems. By incorporating food sovereignty principles into their culinary practices, chefs can contribute to the preservation of diverse food traditions and support sustainable food systems.

How Chefs Can Use Culinary Anthropology to Deepen

Cultural Connections and Enhance Their Practice

For chefs, culinary anthropology is not just an academic field but a source of inspiration and ethical guidance. By incorporating the principles of culinary anthropology into their work, chefs can create dishes that go beyond flavour to convey cultural meaning and connect diners with the broader human story behind the food. This approach invites chefs to view each dish as a medium of cultural exchange, where flavours, ingredients, and techniques reveal insights about the people and places that inspired them.

One way chefs can use culinary anthropology is by incorporating authentic ingredients and preparation methods that honour the origins of a dish. For instance, a chef preparing a traditional Moroccan tagine might choose to use preserved lemons and olives sourced from North Africa, acknowledging the regional specificity of these ingredients. Similarly, a chef creating a Japanese-inspired menu might focus on seasonal ingredients and incorporate Japanese cooking techniques, such as umami balancing and knife skills, to honour the cultural principles of washoku, or the Japanese culinary philosophy of harmony. By using ingredients and techniques that reflect a dish's cultural origins, chefs can create a dining experience that resonates with authenticity and respect.

Culinary anthropology also encourages chefs to be mindful of the social and environmental contexts of the foods they use. For example, understanding the significance of communal eating in Mediterranean cultures can inspire chefs to create shared dining experiences, such as mezze or tapas, that bring people together in a spirit of hospitality and community. Similarly, an appreciation of the agricultural practices behind traditional ingredients can guide chefs in sourcing responsibly, whether by choosing locally grown produce, supporting fair-trade suppliers, or working with farms that prioritize sustainable practices.

Innovation, too, can benefit from the insights of culinary anthropology. Chefs often face the challenge of balancing tradition with creativity, and culinary anthropology offers a framework for respectful fusion. By understanding the historical and cultural contexts of the foods they are blending, chefs can create new dishes that pay homage to both traditions without erasing or diminishing either. For instance, a chef might combine the techniques of French pâtisserie with the flavours of Indian desserts, creating a fusion dessert that respects the culinary principles of both cultures. Culinary anthropology provides chefs with the tools to innovate thoughtfully, creating dishes that celebrate diversity while honouring the roots of each cultural influence.

Furthermore, culinary anthropology encourages chefs to engage with the cultural narratives surrounding food, providing diners with a deeper and more enriching dining experience. By sharing the stories of the ingredients, dishes, and techniques they use, chefs can create a dining atmosphere that fosters curiosity, understanding, and appreciation. This storytelling aspect of culinary anthropology transforms the dining experience into an educational journey, allowing chefs to play a role in cultural preservation and dialogue.

For example, a chef introducing a Peruvian-inspired dish might explain the historical significance of quinoa as a staple grain in Andean culture, or highlight the health benefits of traditional ingredients like aji peppers. These narratives offer diners a window into another culture's relationship with food, enriching their understanding and enhancing their enjoyment of the meal.

Culinary anthropology offers chefs a powerful way to connect with the foods they work with and to honour the diverse culinary traditions that shape global cuisine. By viewing food as a cultural artifact, chefs can deepen their understanding of the world's food heritage, create dishes that honour and respect cultural diversity, and foster a dining experience that celebrates the universal language of food. Culinary anthropology empowers chefs to approach their craft with greater mindfulness and sensitivity, enriching both their personal practice and the experiences they offer to diners.

CHAPTER 2: THE EVOLUTION OF GASTRONOMY

Early Food Traditions And The Origins Of Culinary Practices In Ancient Civilizations

The origins of culinary practices trace back to early human societies, where food preparation emerged as both a survival strategy and a means of social connection. In the earliest communities, gathering, hunting, and foraging provided sustenance, but as humans learned to harness fire, the practice of cooking evolved, transforming raw ingredients into new forms that offered better taste, digestibility, and nutrition. Fire allowed early humans to experiment with foods that were previously inedible or dangerous, paving the way for basic techniques such as roasting, drying, and boiling. This innovation marked the beginning of gastronomy, as communities began to recognize that food was more than mere sustenance—it could be crafted, manipulated, and enjoyed.

Archaeological evidence from prehistoric sites

indicates that early humans had a surprisingly sophisticated approach to food preparation. Stone tools found near ancient hearths suggest that early hunter-gatherers practiced rudimentary butchery techniques to maximize the yield from hunted animals. Similarly, the discovery of grinding stones and cooking pits points to the processing of grains and plant roots, demonstrating an early understanding of flavour and texture enhancement. For chefs today, these discoveries offer a glimpse into humanity's primal relationship with food and an appreciation for the ingenuity that defined early culinary practices.

As human societies grew more complex, the development of agriculture transformed food culture, enabling communities to settle and cultivate crops. This agricultural revolution, which began around 10,000 BCE in the Fertile Crescent, changed the course of human history, giving rise to stable food sources and allowing for surplus production. Wheat, barley, and legumes became staple crops, enabling societies to sustain larger populations and establish the first complex civilizations. For the first time, food storage became essential, leading to innovations such as fermentation, drying, and pickling. These practices, which allowed for the preservation of surplus crops, laid the groundwork for gastronomy by encouraging experimentation with flavour and texture.

The domestication of animals further contributed to early culinary practices, providing access to dairy products, meat, and additional sources of protein. Cattle, sheep, and goats became integral to the diets of ancient societies, influencing culinary techniques and dietary preferences that remain prevalent in many cultures today. The transition from hunter-gatherer communities to agrarian societies marked a turning point in culinary history, fostering the development of complex cooking methods and setting the stage for the diverse culinary traditions that define global gastronomy today.

Key Developments in Cooking Techniques and Flavour Profiles Across History

As civilizations evolved, so did their culinary techniques and approaches to flavour. Ancient Egyptians, for example, were known for their use of herbs and spices to enhance the flavour of their dishes. They cultivated and traded various aromatic plants, including coriander, cumin, and anise, which they used not only for cooking but also in religious ceremonies and medicinal practices. The introduction of spices to the Egyptian diet reflects an early appreciation for complex flavour profiles, showcasing a sophisticated approach to seasoning that influenced later civilizations across the Mediterranean.

In Mesopotamia, considered one of the cradles of civilization, early texts reveal detailed recipes and cooking methods that demonstrate an advanced culinary tradition. The oldest known cookbook, the Akkadian tablets from the Mesopotamian city of Mari (dating to around 1750 BCE), contains recipes for stews, breads, and other dishes that required skilled preparation and a nuanced understanding of flavour. The recipes call for ingredients such as onions, garlic, leeks, and meats, seasoned with herbs and spices, illustrating a deliberate approach to balancing flavours. These tablets highlight the importance of culinary knowledge in Mesopotamian society, where cooking was an esteemed craft, and chefs were respected artisans.

Ancient Greece also played a significant role in the evolution of gastronomy, with philosophers and poets often discussing food as an art form. The Greeks were among the first to develop a philosophy of gastronomy, believing that food should satisfy both body and soul. Influential Greek thinkers, such as Archestratus, often celebrated the pleasures of dining, advocating for the use of high-quality ingredients and emphasizing moderation and balance in meals. Greek cooking techniques included grilling, boiling, and baking, and the Greeks were known for their love of olives, honey, fish, and grains. Their culinary practices, particularly their focus on harmony and simplicity, laid the foundation for Mediterranean cuisine and contributed to the later development of Western culinary traditions.

The Roman Empire further advanced culinary practices, drawing influences from the various cultures they encountered across their vast territories. Romans adopted culinary techniques and ingredients from Greece, North Africa, and the Middle East, creating a fusion cuisine that reflected the empire's multicultural makeup. Roman chefs used complex cooking methods, such as baking in clay ovens and fermenting fish sauce (garum), which became a prized condiment. Roman banquets were elaborate affairs, with multiple courses and dishes that demonstrated both culinary skill and social status. The Romans also placed a strong emphasis on flavour, using spices, herbs, and sweeteners such as honey to create sophisticated dishes that were enjoyed by the elite. This Roman approach to gastronomy, with its focus on elaborate meals and diverse flavours, exemplifies the evolving complexity of culinary practices in ancient civilizations.

The spice trade, which connected Asia, the Middle East, and Europe, played a crucial role in shaping global cuisine by introducing exotic flavours to new regions. Spices like cinnamon, black pepper, and cloves were highly sought after, becoming symbols of wealth and power. The trade routes facilitated not only the exchange of ingredients but also the diffusion of cooking techniques, leading to the early forms of culinary fusion that continue to define gastronomy today. For chefs, understanding these historical developments provides context for the diverse flavours and techniques available in modern cooking, reminding them that culinary innovation has always been driven by cross-cultural exchanges.

The Role of Food in Cultural Rituals and Ceremonies, with Examples from Mesopotamia, Egypt, and Ancient Greece

Food has long played a central role in cultural rituals and ceremonies, serving as a medium through which societies express their beliefs, honour their deities, and mark significant life events. In ancient Mesopotamia, food offerings were integral to religious rituals, with temples often receiving donations of bread, beer, and meat to appease the gods. The Mesopotamians believed that the gods required nourishment, and they viewed food as a means of maintaining cosmic balance. Large feasts were organized to celebrate religious festivals, during which both the elite and commoners would share in communal meals, reinforcing social bonds and ensuring divine favour. These feasts were not merely acts of consumption but were imbued with symbolic significance, reflecting the Mesopotamians' belief in the interconnectedness of food, spirituality, and community.

In ancient Egypt, food also held deep religious and symbolic meaning, with offerings of bread, beer, fruits, and meats commonly placed in tombs to ensure the deceased had sustenance in the afterlife. The Egyptians viewed food as a bridge between the earthly and divine realms, and many of their rituals centred around feeding both the living and the dead. Pharaohs and nobles would commission elaborate tombs filled with food offerings and kitchenware, believing that these provisions would secure their place in the afterlife. This practice highlights the Egyptians' profound reverence for food and its role in their cosmology, illustrating how culinary practices were deeply intertwined with their religious beliefs.

In ancient Greece, food was central to communal festivals and ceremonies that honoured the gods and celebrated civic unity. The symposium, a gathering of Greek men for philosophical discussions and entertainment, was as much about food as it was about intellectual discourse. During these gatherings, participants would share wine and small bites while engaging in debates, recitations, and socializing, reinforcing the importance of food in fostering intellectual and social bonds. Greek religious festivals, such as the Panathenaea in Athens, featured sacrificial offerings of livestock, with the meat distributed among the community after the ritual. This sharing of food reinforced social cohesion and reflected the Greeks' belief in the sanctity of communal meals.

The ancient Romans also incorporated food into their religious practices, using it to celebrate festivals dedicated to their gods. The Saturnalia festival, for instance, involved days of feasting, gift-giving, and social role reversals, where slaves were temporarily freed and allowed to dine with their masters. This festival highlighted the Roman appreciation for food as both a physical and social experience, blurring the boundaries between different classes and reinforcing social bonds. Roman funeral feasts, known as "convivia," honoured the deceased and allowed the community to come together in mourning and remembrance. The Romans believed that sharing a meal in memory of the deceased would honour their life and provide closure for the family. Through these practices, the Romans demonstrated the symbolic power of food in expressing grief, celebration, and unity.

Each of these civilizations viewed food not merely as sustenance but as an integral part of their cultural and spiritual identity. The rituals, ceremonies, and festivals surrounding food served to reinforce social hierarchies, build community bonds, and honour the divine. For modern chefs, these ancient practices offer inspiration to view food as more than just a culinary experience. By understanding the historical and cultural significance of food in rituals, chefs can approach their craft with a deeper awareness of food's role in fostering human connection and celebrating shared values.

These ancient traditions remind chefs of the potential for food to convey messages of reverence, unity, and respect across cultures. Whether through offering a simple meal to commemorate an event or designing a tasting menu that tells a story, chefs today can draw from these traditions to create dining experiences that transcend taste and touch upon the cultural, spiritual, and emotional facets of food.

CHAPTER 3: FOOD AS CULTURAL IDENTITY

Food As A Reflection Of Cultural Identity And Regional Pride

Food is one of the most powerful expressions of cultural identity. Each dish, ingredient, and flavour often reflects a rich tapestry of history, geography, and social customs, making cuisine an essential component of regional pride. When people gather around traditional foods, they share not only a meal but also a collective memory—a taste of their shared history and values. For centuries, communities have used food to assert their identity, celebrate their heritage, and distinguish themselves from others. From spices carefully cultivated in India to indigenous maize varieties in Mexico, ingredients and cooking methods become symbols of place, carrying the essence of a region's landscape, climate, and traditions.

Culinary practices shape and reflect identity by capturing the unique stories of a people, passed down through generations. In regions where food remains closely tied to tradition, specific flavours, ingredients, and cooking techniques become central to how people define themselves. Food is also deeply entwined with regional pride, especially in areas where culinary practices are fiercely guarded and celebrated. For example, the French hold their cuisine in high regard, often viewing the preservation of their culinary techniques as part of their national identity. Similarly, Japanese culinary traditions—such as the emphasis on seasonal ingredients, minimalism, and balance— are seen as extensions of Japanese philosophy and aesthetics. In many cultures, cooking techniques and ingredients become symbolic markers, encapsulating the natural resources and historical context of the region they represent.

Regional dishes often contain specific ingredients that are unique to the landscape and agricultural practices of the area. For instance, the Mediterranean region's reliance on olive oil, tomatoes, and herbs is a direct reflection of its climate, which supports the growth of these ingredients. Similarly, Scandinavian cuisines are characterized by preservation techniques like pickling and fermenting, developed to endure long, harsh winters when fresh produce was scarce. By understanding how regional pride shapes food culture, chefs can better appreciate the relationship between place, cuisine, and identity. Each dish can be seen as a map of sorts, illustrating the physical and cultural landscapes of its origin.

Moreover, food serves as a medium of cultural expression, particularly for marginalized or diasporic communities who use traditional recipes to maintain a connection with their heritage. Immigrant communities, for example, often bring their culinary practices with them, creating pockets of cultural identity in their adopted countries. These foods provide comfort and familiarity, allowing individuals to stay connected to their roots while navigating life in a new environment. The emergence of "Little Italy" in cities across the United States or "Chinatown" districts worldwide illustrates how food becomes a foundation of cultural identity, preserving language, values, and customs for future generations.

As chefs, understanding food's role in cultural identity provides an opportunity to honour the diversity and richness of global culinary practices. By approaching food with respect for its cultural significance, chefs can participate in a culinary tradition that celebrates, rather than dilutes, regional pride. Whether they are preparing a dish exactly as it is made in its homeland or innovating while preserving its spirit, chefs have the power to create connections through food, bridging cultures and celebrating identity.

Case Studies of Culinary Identity in Italy, India, and Mexico

Each culture has unique culinary traditions that embody its history, environment, and values. Examining specific cases from Italy, India, and Mexico reveals how food serves as an integral part of cultural identity, showcasing the unique flavours and philosophies that define each region's cuisine.

In Italy, culinary identity is strongly tied to local, seasonal ingredients and centuries-old techniques passed down within families and communities. Italian cuisine places immense value on simplicity and quality, with regional pride evident in dishes like Neapolitan pizza, Bolognese pasta, and Tuscan olive oil. Each Italian region boasts its specialties, influenced by the local terrain and climate. For example, the coastal region of Liguria uses a rich variety of seafood, while the mountainous region of Piedmont is known for truffles and game meats. Italian cuisine is less about elaborate techniques and more about allowing high-quality ingredients to shine, a reflection of Italy's agrarian roots and deep connection to its land. Chefs who understand Italian culinary identity respect this simplicity, focusing on fresh ingredients and traditional methods to honour Italy's food heritage.

In India, food reflects a rich history of cultural diversity, with flavours and techniques influenced by centuries of trade, migration, and colonization. India's cuisine is renowned for its complex use of spices, each carefully selected for its flavour, aroma, and health benefits. Regional diversity is also significant; for instance, North Indian cuisine includes wheat-based breads and rich, spiced gravies, while South Indian cuisine is characterized by rice, coconut, and a focus on sour and tangy flavours. The cultural diversity within India is evident in its vegetarian traditions, influenced by Hinduism, Buddhism, and Jainism, which emphasize non-violence and respect for all living beings. In India, food is a vehicle for cultural and religious expression, with many meals prepared as offerings in rituals and ceremonies. For chefs, understanding the significance of spices, regional variations, and religious practices in Indian cuisine is essential to honouring its cultural identity.

In Mexico, food is both a celebration of indigenous ingredients and a testament to resilience in the face of colonial influence. Mexican cuisine is rooted in ingredients such as corn, beans, chilies, and tomatoes, staples that have sustained indigenous populations for centuries. The preparation of dishes like tamales, mole, and tortillas reflects traditional techniques, often involving grinding, slow-cooking, and layering flavours over time. Mexican food is known for its bold flavours and use of native ingredients, yet it also bears the marks of Spanish influence, particularly in its use of meats and dairy. Mexican cuisine is a source of national pride, representing a blend of indigenous and colonial elements. For chefs, understanding the resilience and creativity behind Mexican dishes can inspire respectful and authentic culinary practices.

Through these examples, we see how each culture uses food to express its unique values, history, and environment. Italian cuisine honours the land, Indian cuisine reflects diversity and spirituality, and Mexican cuisine celebrates resilience and indigenous pride. For chefs, these case studies illustrate the depth of cultural identity embedded in food, offering guidance on how to approach these traditions with respect and integrity.

How Chefs Can Honour and Preserve Culinary Traditions While Innovating

For chefs, the challenge of honouring culinary traditions while embracing innovation requires both respect and creativity. As the global culinary landscape evolves, chefs are increasingly expected to blend tradition with new techniques, creating dishes that pay homage to their origins while meeting contemporary tastes. This balancing act involves an understanding of cultural values and a commitment to authenticity, ensuring that the essence of traditional dishes is preserved even when adapted to modern contexts.

One way chefs can honour culinary traditions is by sourcing authentic ingredients and using traditional techniques. For example, a chef preparing a Japanese-inspired dish may choose to use dashi—a traditional Japanese broth made from kombu and bonito flakes—rather than a generic broth, thereby maintaining the authentic umami flavours central to Japanese cuisine. Similarly, when making Italian pasta, using semolina flour and practicing traditional pasta-shaping methods shows respect for the Italian culinary heritage. Chefs who incorporate authentic ingredients and techniques demonstrate a dedication to preserving the cultural essence of their dishes, offering diners an experience that is both authentic and respectful.

In addition to ingredients and techniques, chefs can engage with the stories and philosophies behind traditional foods, bringing cultural context to their creations. Sharing the history or significance of a dish on a menu, for instance, can transform a meal into an educational experience for diners, fostering appreciation for the culinary heritage being represented. This approach allows chefs to bridge cultures and educate diners on the cultural importance of the foods they enjoy. For example, a chef serving mole, a complex Mexican sauce with dozens of ingredients, might explain its ceremonial origins and the care that goes into its preparation, inviting diners to appreciate the labour and love embedded in traditional Mexican cooking.

Innovation is inevitable in any living tradition, and culinary anthropology encourages chefs to experiment within respectful boundaries. Fusion cuisine, for instance, can be a respectful form of innovation when done thoughtfully, drawing from multiple traditions while honouring each source. A chef might experiment with Indian spices in a French-style dish, combining the techniques of French cuisine with the bold flavours of India. The key is to maintain balance and avoid distilling the dish to a superficial level, ensuring that the flavours, techniques, and cultural meanings are preserved. Chefs who innovate in this way contribute to the evolution of culinary traditions, adding new layers to cultural dishes without compromising their integrity.

However, as chefs innovate, it's essential to avoid cultural appropriation, which reduces cultural elements to trends or stereotypes. Cultural appropriation often occurs when a dish is stripped of its cultural significance and presented without proper context, ignoring the history and identity it represents. For instance, taking a traditional Mexican street food and rebranding it without recognizing its cultural roots can be seen as a form of disrespect. To avoid this, chefs should educate themselves about the origins of the dishes they reinterpret, collaborate with individuals from those cultures, and remain transparent about the inspirations behind their innovations.

Chefs can honour and preserve culinary traditions by respecting the cultural identity embedded in each dish. Through thoughtful sourcing, storytelling, and innovation, chefs have the power to bring cultural richness to their menus, offering diners a deeper appreciation for the diversity of global cuisine. By embracing culinary anthropology, chefs can craft dishes that not only delight the palate but also celebrate and preserve the cultural identities that make each cuisine unique.

CHAPTER 4: RITUALS AND FOOD IN RELIGIOUS PRACTICES

Historical Context Of Food In Religious Rituals (E.g., Feasts, Fasting)

Throughout history, food has held a significant place in religious rituals and practices across cultures. The act of sharing food has been a means of connecting with the divine, expressing devotion, and fostering community bonds. Across different religions, food serves as a tangible representation of intangible beliefs, symbolizing gratitude, sacrifice, and reverence. From the lavish feasts in honour of gods and saints to periods of fasting meant for self-purification and reflection, food has always been an integral part of religious life.

Religious feasts and celebrations often include specific

dishes, ingredients, and preparation methods that are imbued with symbolic meaning. In ancient societies, sacrificial offerings of food were common practices meant to appease the gods and bring blessings upon the community. The Egyptians, for instance, made elaborate food offerings to their deities, believing that these would sustain the gods and secure divine favour. Similarly, the Greeks and Romans organized grand feasts, often as public gatherings, to honour their pantheon of gods. The foods used in these offerings—such as wine, grains, and fruits—were chosen for their abundance and represented fertility, prosperity, and abundance.

Fasting, too, is a practice found in many religious traditions, symbolizing purification and spiritual discipline. Ancient cultures viewed fasting as a way to demonstrate devotion, achieve spiritual clarity, and detach from worldly desires. In Hinduism, fasting has been practiced for thousands of years as a means of focusing the mind and body on spiritual growth. Fasting periods in many religions, such as Lent in Christianity and Yom Kippur in Judaism, serve as times for reflection and repentance, creating a sense of renewal through abstaining from certain foods. These periods are often marked by both physical and spiritual introspection, underscoring the belief that limiting one's diet can lead to heightened spiritual awareness.

The significance of food in religious rituals demonstrates the profound connection between nourishment and spirituality. Food is not simply seen as a physical sustenance but as a gift from a higher power, meant to be shared with gratitude. For chefs, understanding this historical context is essential in recognizing the sacred relationship between food and religion. The dishes they prepare can be a form of respect and acknowledgment, honouring the traditions that infuse each meal with cultural and spiritual significance.

Examination of Food in Religious Ceremonies Across Christianity, Islam, Hinduism, and Indigenous Beliefs

Food practices vary widely across religions, but they all share a common theme: the use of food as a means of spiritual expression and community building. Exploring the role of food in Christianity, Islam, Hinduism, and Indigenous beliefs provides valuable insights into how chefs can respect these traditions and create dining experiences that honour diverse spiritual practices.

In Christianity, food plays a significant role in worship and community. The most notable example is the Eucharist or Holy Communion, a sacrament practiced by many Christian denominations in which bread and wine are consecrated and consumed in remembrance of Jesus Christ's Last Supper with his disciples. The bread symbolizes Christ's body, and the wine represents his blood, making this ritual an act of spiritual unity and remembrance. Different Christian denominations have variations on the Eucharist, with some using unleavened bread and others preferring wine or grape juice. Beyond the Eucharist, Christian celebrations such as Christmas and Easter often include special meals and dishes that reflect cultural traditions within the faith. Christmas feasts, for example, are occasions for families to gather, and in many cultures, specific dishes—like roast turkey, ham, or plum pudding—carry symbolic meaning. Chefs preparing meals for Christian clientele may consider the sacred nature of the Eucharist and the cultural symbolism of traditional holiday dishes, treating these occasions with the reverence they deserve.

Islam also incorporates food deeply into its religious practices, with dietary laws and fasting as central components of worship. The concept of halal (permissible) governs the foods Muslims are allowed to consume, requiring that meat be prepared through a ritualistic slaughter that emphasizes respect for the animal. During Ramadan, the ninth month of the Islamic lunar calendar, Muslims observe a daily fast from sunrise to sunset, refraining from eating, drinking, and other physical needs as a form of self-discipline and spiritual reflection. The fast is broken each evening with an iftar meal, often beginning with dates and water, followed by a variety of dishes that reflect the cultural diversity of the Islamic world. Celebrating Eid al-Fitr, the festival that marks the end of Ramadan, is a joyful occasion marked by feasting, where families and communities gather to enjoy foods like kebabs, sweets, and other traditional delicacies. Chefs catering to Muslim clients can respect these traditions by ensuring that meals are halal-compliant and by offering culturally relevant dishes during Ramadan and Eid celebrations, fostering an inclusive dining experience.

Hinduism views food as a means of connecting with the

divine, with a strong emphasis on purity and respect for all living beings. Many Hindus adhere to a vegetarian diet, guided by the principle of ahimsa, or non-violence, which discourages the consumption of meat as a way to show respect for all forms of life. In Hindu temples, offerings known as prasad are made to deities, with the food later distributed to devotees as a blessed item. These offerings often include fruits, sweets, and milk-based dishes, symbolizing abundance and purity. Certain Hindu festivals, like Diwali, are marked by elaborate feasts and specific sweets, each with regional variations across India. Fasting is also a common practice in Hinduism, with different fasts observed on specific days or during festivals, symbolizing devotion and self-discipline. For chefs, understanding Hindu dietary restrictions and the cultural meanings of certain foods can guide respectful menu planning, especially when catering for Hindu clients who may have specific preferences for vegetarian or vegan options during religious observances.

Indigenous beliefs around the world demonstrate a profound respect for nature, viewing food as a gift from the earth that must be honoured and shared responsibly. Many Indigenous cultures practice ceremonies around the harvest and hunting seasons, with food offerings made to spirits or ancestors in gratitude for the bounty of nature. For example, Native American tribes in North America celebrate the harvest with feasts that include corn, beans, and squash—often called the "Three Sisters," these crops hold cultural significance as staples that sustain the community. In the Andean regions of South America, Indigenous communities honour Pachamama, or

Mother Earth, with food offerings that acknowledge her role in providing sustenance. These ceremonies often involve ritualistic cooking and sharing, emphasizing communal responsibility and a deep spiritual connection to the land. Chefs working with Indigenous foods or preparing meals for Indigenous clients can honour these practices by sourcing local ingredients, practicing sustainable cooking, and acknowledging the spiritual significance of certain foods.

Across these religions, food serves as a conduit for expressing devotion, showing gratitude, and fostering community. By understanding these diverse culinary traditions, chefs can approach their work with greater sensitivity and insight, creating menus and dining experiences that respect the values and beliefs of people from different faiths.

Practical Application for Chefs in Creating Culturally Sensitive Dishes for Diverse Clientele

For chefs, creating culturally sensitive dishes involves more than simply following dietary restrictions— it requires a deep understanding of the symbolic meanings and spiritual values associated with food. To cater to clients from different religious backgrounds, chefs can take practical steps to ensure that their menus honour these traditions respectfully.

One essential consideration for chefs is understanding dietary laws specific to each religion. In addition to familiarizing themselves with halal and kosher dietary guidelines, chefs should be aware of vegetarian preferences in Hinduism and the importance of specific fasting practices. Respecting these dietary laws demonstrates a commitment to inclusivity, ensuring that clients can dine without compromising their beliefs. For instance, a chef preparing a meal for a Muslim client may ensure that all meat is halal-certified, while a meal for Jewish clients could follow kosher guidelines, including separate preparation areas for meat and dairy. This attention to detail not only shows respect for religious practices but also builds trust with clients, as they know their beliefs are being honoured.

Ingredient sourcing is another important aspect of creating culturally respectful dishes. Certain ingredients hold special meaning in various religions, and chefs can enhance the authenticity of their dishes by sourcing these ingredients carefully. For example, dates are significant in Islam, particularly during Ramadan when they are eaten to break the fast, so sourcing high-quality dates adds cultural depth to the meal. In Hinduism, the use of pure ingredients, such as ghee and milk, is important in offerings made to deities. For chefs preparing Indian-inspired dishes for Hindu clients, using authentic ingredients can enhance the spiritual value of the meal. Sourcing locally and sustainably can also show respect for Indigenous beliefs, aligning with their view of food as a sacred gift from the earth.

Mindfulness in menu design is essential when serving religiously diverse clientele. Offering a variety of vegetarian and vegan options can cater to clients from Hindu, Buddhist, and Jain traditions, while labelling halal and kosher options allows Muslim and Jewish clients to make informed choices. Chefs can also design menus that reflect the spirit of religious observances. For example, a chef preparing an iftar meal during Ramadan could create a menu that begins with dates and includes light, nourishing dishes to break the fast. During Lent, a Christian observance involving dietary restrictions, chefs could offer fish-based dishes as an alternative to meat, acknowledging the cultural significance of these choices.

Beyond dietary guidelines, creating a culturally respectful dining atmosphere is crucial. This might involve explaining the cultural significance of certain dishes or ingredients, helping diners appreciate the heritage behind their meals. By including brief descriptions of the cultural importance of each dish on the menu, chefs can provide an educational experience that fosters understanding and respect. For instance, a chef serving a Native American-inspired dish might explain the role of corn, beans, and squash in Indigenous cuisine, giving diners insight into the cultural and agricultural practices behind the ingredients.

Respectful fusion and innovation also play a role in culturally sensitive dining. While it is common for chefs to experiment with fusion cuisine, it's essential to approach religious and cultural dishes with sensitivity, ensuring that the integrity of the original dish is preserved. For example, when creating a fusion dish inspired by Indian or Mexican cuisine, chefs might avoid mixing incompatible flavours or presenting traditional foods in ways that could be perceived as disrespectful. Consulting with individuals from those cultural backgrounds, or researching traditional preparation methods, can help chefs innovate while maintaining authenticity and respect.

Creating culturally sensitive dishes for a diverse clientele requires a commitment to understanding the religious and cultural values that influence people's relationships with food. By honouring dietary laws, sourcing ingredients mindfully, designing inclusive menus, fostering an educational dining experience, and practicing respectful fusion, chefs can create meals that not only satisfy the palate but also honour the spiritual and cultural significance of food. For chefs, this approach enhances their role as cultural ambassadors, allowing them to bridge culinary traditions and celebrate the diversity of human beliefs through the universal language of food.

CHAPTER 5: FOOD AND SOCIAL HIERARCHY

How Food Reflects Social Status And Hierarchy In Various Societies

Food has long been a marker of social status and hierarchy, reflecting the distinctions between different classes within societies. From the ingredients used to the manner of serving and the rituals surrounding meals, food communicates messages about power, wealth, and privilege. Across cultures and historical periods, what and how people ate was often a direct indicator of their place in society, with certain foods, preparation methods, and dining etiquettes reserved for the elite, while others symbolized simplicity or humility associated with lower classes.

In many societies, access to specific foods has been a privilege of the upper classes. Luxurious and rare ingredients, such as spices, fine wines, and exotic meats, were symbols of wealth and power. Historically, spices like saffron, cinnamon, and cloves were worth their weight in gold, accessible only to those who could afford them. Similarly, meat, particularly from exotic animals, was considered a luxury and often featured in feasts to demonstrate the wealth of the host. The lower classes, on the other hand, typically subsisted on simpler, more readily available foods such as grains, legumes, and locally sourced vegetables. These dietary distinctions reinforced social divisions, as what one ate was not only a matter of sustenance but also a symbol of one's social position.

Food preparation and presentation also serve as indicators of social hierarchy. In many aristocratic households and royal courts, elaborate multi-course meals were prepared by skilled chefs and served by attendants, highlighting the host's wealth and refinement. The complexity of the dishes and the precision of their presentation were not merely for taste but were symbols of social prestige. Table settings, elaborate service styles, and lavish displays of food signified a level of sophistication and exclusivity available only to the elite. By contrast, the common classes typically shared meals in simpler settings, with communal dishes that fostered unity and cooperation rather than hierarchy. For example, in rural communities across Europe, communal stews and bread, often cooked over shared fires, were staples of daily life, representing egalitarianism rather than social stratification.

Even the times and places where people ate reinforced social distinctions. In some cultures, dining rooms or feasting halls were reserved for the privileged, while others dined in kitchens or outdoor spaces. The timing of meals, too, was indicative of social rank; nobility might enjoy leisurely, drawn-out feasts, while labourers and peasants had limited time to eat during the day. These distinctions in dining practices created a visible divide between the elite and the working classes, as they symbolized not only the food itself but the luxury of time, space, and resources that only the wealthy could afford.

The concept of food as a social marker continues today, albeit in more subtle forms. Fine dining, artisanal ingredients, and organic or specialty foods are often associated with higher social status, while fast food and processed meals are more affordable but carry a perception of being less desirable. The rise of food trends and "foodie" culture has also contributed to new forms of culinary elitism, where knowledge of niche ingredients or exclusive dining experiences becomes a means of distinguishing oneself socially. For chefs, understanding this dynamic offers insight into the cultural meanings that diners may associate with certain foods and dining experiences, allowing them to create meals that consider not only taste but the deeper social and symbolic connotations of food.

Historical Examples from Royal Banquets to Communal Meals

Throughout history, royal banquets, feasts, and communal meals have highlighted the interplay between food and social hierarchy. These events illustrate how food served not only to nourish but also to communicate power dynamics, reinforce social structures, and celebrate communal identity. By examining these examples, chefs can gain insight into the historical context of dining practices and the ways in which food has traditionally signified status.

In ancient Egypt, royal banquets held by the pharaohs were grand affairs meant to demonstrate wealth, power, and divine favour. These banquets featured an array of meats, fruits, breads, and wines, all prepared with ingredients and spices that were rare and expensive. The pharaoh and his closest attendants would dine on rich dishes, while lesser nobles and servants ate simpler fare, seated at a distance from the royal table. This separation in the type and quality of food served reinforced the social hierarchy, with the most privileged individuals enjoying the finest offerings, while those of lower status received more modest portions. The grandeur of the banquet itself, with golden utensils, luxurious table settings, and elaborate presentations, reinforced the message of the pharaoh's supreme status.

Similarly, the courts of medieval Europe were known for their extravagant feasts, where food was used to showcase the wealth and power of the nobility. At these banquets, whole roasted animals, often adorned with gold leaf, were presented as centrepieces, alongside pies filled with exotic birds and towers of fruit. The elaborate preparations were designed to impress guests and assert the host's social dominance. These feasts followed strict etiquette, with seating arrangements reflecting the social hierarchy and only those of higher rank permitted to partake in certain dishes. For instance, in English and French courts, only the most esteemed guests were served game meats, while commoners were offered bread and broth. Feasts like these were highly theatrical, with musicians, dancers, and jesters providing entertainment, all reinforcing the host's wealth and influence.

Conversely, communal meals in many Indigenous cultures around the world emphasize a more egalitarian approach to food, reflecting collective values rather than hierarchical structures. Among Native American tribes, communal feasts celebrated harvests, honoured tribal leaders, and marked spiritual ceremonies, where food was shared equally among members. In the Andean communities of South America, the tradition of "minka"—a form of collective labour followed by shared meals—reinforced social unity and reciprocity. These communal meals were less about asserting individual status and more about fostering community bonds. Everyone contributed to the feast, and everyone was entitled to an equal share, underscoring values of cooperation and respect.

In Japan, the tea ceremony, or "chanoyu," reflects a blend of hierarchy and humility in its approach to food and drink. While the ceremony is highly structured, emphasizing discipline, order, and reverence, it does not focus on lavish display or material wealth. Instead, the tea ceremony highlights aesthetic appreciation, mindfulness, and respect for the natural world. The host, often of higher social status, prepares tea for guests, embodying a spirit of humility and service. The tea ceremony serves as an equalizer, where individuals from different social backgrounds come together in an atmosphere of harmony, simplicity, and shared purpose.

These historical examples reveal the diverse ways in which societies have used food to express social hierarchies or foster communal bonds. By studying these traditions, chefs can gain insight into the symbolic meanings associated with different dining experiences and consider how they might incorporate elements of hierarchy or unity into their own culinary offerings.

Insights for Chefs on Creating Experiences That Respect Diverse Social Backgrounds

For chefs, understanding the ways in which food reflects social hierarchy and cultural values provides valuable insight into how they can create dining experiences that respect and honour the backgrounds of their clients. Crafting a meal that considers the social dynamics of food can help chefs engage with the diversity of their clientele, creating an atmosphere that feels inclusive and respectful of various cultural traditions.

One practical approach is customizing the dining experience to cater to different social preferences. Chefs can design menus with options that cater to both communal and individual styles of dining, allowing guests to choose an experience that aligns with their cultural and social values. For example, offering family-style dining with large platters and shared dishes can foster a sense of unity, appealing to clients from cultures that value communal meals. Conversely, for clients who may associate individual servings with elegance and formality, chefs can design multi-course meals with individually plated dishes, reflecting a sense of refinement and exclusivity. By offering a range of dining formats, chefs can respect and accommodate the diverse preferences of their guests.

Elevating simple ingredients is another way for chefs to create a sense of inclusivity while respecting diverse social backgrounds. While fine dining often emphasizes luxurious and rare ingredients, there is also an art in transforming simple, everyday ingredients into exquisite dishes. This approach reflects the values of many traditional cultures, where common ingredients such as rice, beans, and vegetables are treated with care and creativity. By elevating humble ingredients, chefs demonstrate that delicious and beautiful food is not exclusive to the elite. This practice can be particularly meaningful for clients who come from modest backgrounds, as it emphasizes the beauty and value in accessible ingredients, making the dining experience more relatable and inclusive.

Incorporating cultural rituals and customs into the dining experience can also help chefs create a sense of respect for diverse social backgrounds. For example, offering a small, ceremonial dish at the beginning of the meal, such as a blessing bowl with grains or a seasonal tea, can set a respectful tone and invite diners to reflect on the cultural origins of the meal. Similarly, explaining the cultural significance of certain dishes or ingredients can enhance the dining experience, helping guests appreciate the social and historical contexts behind each plate. Chefs might, for instance, share the story of a dish inspired by Indigenous cuisine, or explain the symbolism of saffron in Persian cooking. These narratives not only add depth to the meal but also show a respect for the social history embedded in culinary traditions.

In modern dining, where "foodie" culture has contributed to culinary elitism, chefs can also practice culinary humility by avoiding pretension and embracing diversity in their menus. This involves creating a balance between fine dining elements and accessible flavours, ensuring that the menu resonates with a wide audience. For instance, a fine dining restaurant might serve an elevated version of a beloved comfort food, such as macaroni and cheese with artisanal cheeses and truffle oil. This approach respects the nostalgia and familiarity of comfort foods while offering a refined presentation, bridging the gap between luxury and everyday dining. By embracing dishes that feel approachable, chefs can make fine dining more inclusive, appealing to clients from all walks of life.

Finally, mindfulness in pricing and portion sizes is an important consideration for chefs aiming to respect diverse social backgrounds. Offering a range of portion sizes or price points allows guests to choose options that align with their preferences and financial circumstances. Additionally, thoughtful portion sizes can help prevent food waste, a practice that aligns with the values of many cultures where food is viewed as a precious resource. Chefs who approach their menus with mindfulness and inclusivity can create dining experiences that are not only luxurious but also accessible, ensuring that all clients feel welcome and respected.

In summary, by drawing from the historical relationship between food and social hierarchy, chefs can create dining experiences that celebrate diversity, respect cultural values, and foster a sense of unity among diners. Through thoughtful menu design, respect for ingredients, and attention to cultural customs, chefs have the power to transform a meal into an inclusive celebration of both individual and collective identity.

CHAPTER 6: FOOD SCARCITY AND PRESERVATION TECHNIQUES

Historical Challenges Of Food Scarcity And Methods Of Preservation

Throughout human history, food scarcity has been a persistent challenge shaped by seasonal changes, environmental factors, and population growth. In early societies, the availability of food was highly dependent on the local environment, and without modern transportation or preservation technology, communities had to develop methods to manage periods of abundance and scarcity. Seasonal shifts, droughts, floods, and crop failures created unpredictable food supplies, forcing people to find ways to extend the shelf life of their harvests and secure sustenance for leaner times.

In ancient agrarian societies, food scarcity was particularly acute during winter months or dry seasons, when fresh crops were unavailable. The unpredictability of crop yields made it essential for communities to preserve surplus food whenever possible. Traditional agricultural practices included harvesting, drying, and storing grains, which could remain edible for extended periods when properly sealed and protected from moisture. Grains like wheat, rice, and barley became staples in many regions precisely because of their durability and ease of storage. As staple foods that could endure months without spoiling, these grains became cornerstones of both diet and survival.

Hunter-gatherer societies, on the other hand, faced different challenges. Although they were more mobile and could migrate with the seasons to follow food sources, they still needed ways to store and transport food for survival during long journeys or harsh conditions. Techniques such as drying, smoking, and salting meats and fish emerged from the need to make portable, long-lasting sources of protein. For example, the Inuit of the Arctic region used wind-drying and salting methods to preserve fish and seal meat, which were essential food sources during the long, cold winters when fresh food was scarce.

Food preservation also became vital for early civilizations that engaged in trade and exploration. In ancient Egypt, sailors would carry preserved food, often dried or salted fish and bread, to sustain them on long voyages. These preservation techniques allowed for the transport of food across great distances, facilitating trade and exchange between distant regions. By preserving foods, civilizations could establish trade networks, exchanging preserved goods like fish, grains, and spices with neighbouring cultures. Preservation methods thus played a dual role: they were essential for survival during periods of scarcity and also enabled the growth of trade, expanding culinary diversity and cultural interaction.

The need for preservation techniques has remained significant through history, from the medieval period's salted meats and cheeses to the industrial era's innovations in canning and refrigeration. For chefs, understanding these historical challenges sheds light on the resourcefulness of early culinary practices and offers inspiration for using traditional methods to extend the life of food without modern preservatives.

Techniques Like Fermentation, Drying, and Salting Across Cultures

Throughout history, cultures worldwide have developed unique methods for preserving food based on the resources available in their environments. Techniques like fermentation, drying, and salting have become cornerstones of culinary traditions, each with distinct flavours, textures, and nutritional benefits.

Fermentation is one of the oldest and most widely practiced preservation methods, used to convert perishable foods into stable, long-lasting products through the action of beneficial bacteria and yeasts. In East Asia, for instance, fermentation has produced staples such as soy sauce, miso, and kimchi—foods that add depth and complexity to local cuisines. Kimchi, a traditional Korean dish made from fermented cabbage and radishes, is seasoned with salt, chili, and garlic before being fermented in earthenware pots. This process not only preserves the vegetables but also enhances their flavour, creating a dish that is rich in probiotics and vitamins. In Japan, miso, a fermented soybean paste, has been a dietary staple for centuries. Miso adds umami to dishes and offers nutritional benefits, including protein, vitamins, and healthy bacteria, making it both a flavour enhancer and a health food.

In Europe, fermentation techniques were also widely practiced, especially in areas where dairy farming was prevalent. Cheese and yoghurt are products of milk fermentation that have been enjoyed across Europe for centuries. In the Netherlands, France, and Switzerland, cheesemaking became a specialized art, producing varieties with diverse textures, flavours, and aging times. Yoghurt, traditionally consumed in regions like Greece and the Balkans, was valued for its ability to preserve milk and provide a rich source of nutrients. These fermented dairy products became central to the diet and cultural identity of these regions, illustrating how fermentation can adapt to the ingredients and needs of specific climates and societies.

Drying is another ancient preservation method, dating back thousands of years. By removing water from foods, drying inhibits the growth of bacteria and mould, making food safe for extended storage. In arid regions, sun-drying was a natural choice for preserving fruits, vegetables, and meats. Ancient Egyptians were known to sun-dry fruits like dates and figs, which provided a concentrated source of energy and nutrition for long journeys. In the Mediterranean, drying was used to preserve tomatoes, olives, and herbs, all of which are staples in Mediterranean cuisine. Dried tomatoes, for instance, are packed with flavour and can be used to enhance stews, pasta dishes, and breads, adding a concentrated taste of the sun-drenched Mediterranean region to any meal.

In Asia, drying fish and seafood became a widely practiced method, particularly in coastal areas where fresh fish was abundant but difficult to store. In Japan, dried fish products like katsuobushi (dried bonito) are central to the cuisine, providing the base for dashi, a flavourful broth that forms the backbone of many Japanese dishes. In the Philippines, dried fish is a common food staple known as daing, often eaten with rice and used as a topping or ingredient in local dishes. These dried products reflect both the resourcefulness of early preservation methods and the integration of preserved foods into culinary traditions.

Salting, one of the simplest and most effective preservation techniques, has been used globally to extend the shelf life of meats, fish, and vegetables. Salt draws moisture out of foods through osmosis, creating an environment where bacteria cannot thrive. In ancient Rome, salted fish was a common commodity, and the Romans perfected the art of producing garum, a fermented fish sauce made from salted fish that was widely used as a condiment. Garum became a staple across the Roman Empire, illustrating how salting and fermentation techniques were intertwined to create long-lasting, flavourful products.

In Scandinavia, where winters are harsh and fresh

food is scarce, salting became essential for preserving fish, particularly cod and herring. Salted cod, known as klippfisk or bacalhau, became a dietary staple and a culturally significant food that continues to be enjoyed in countries like Norway, Portugal, and Spain. The importance of salted fish is also seen in Jewish cuisine, where salted herring became a beloved delicacy, particularly among Eastern European Jewish communities. For chefs, these examples show how salting not only preserves food but also enhances its flavour, creating foods that are deeply embedded in cultural heritage.

These techniques—fermentation, drying, and salting —have preserved foods across cultures, reflecting the ingenuity of early societies in overcoming the challenge of food scarcity. For modern chefs, these methods offer valuable tools for creating unique, flavourful dishes and a means of connecting with traditional culinary practices.

Application of Preservation Techniques in Modern Culinary Practices

In contemporary cuisine, traditional preservation techniques have experienced a resurgence, not only for their practicality but also for their culinary potential. Modern chefs are increasingly drawn to methods like fermentation, drying, and salting to add complexity, enhance flavours, and reconnect with culinary traditions. By integrating these techniques, chefs can create dishes that resonate with historical depth while appealing to modern tastes.

Fermentation has gained widespread popularity in recent years, celebrated for its ability to transform ingredients into complex, flavour-rich foods. Chefs in fine dining and casual eateries alike are incorporating fermented ingredients into their dishes, often as a way to introduce unique flavours. Fermented foods such as kimchi, sauerkraut, and kombucha have become mainstays on menus, prized for their tangy flavours and probiotic benefits. In addition to traditional fermented items, chefs are experimenting with new ingredients and processes, fermenting everything from fruits to condiments to create innovative flavours. For instance, some chefs are exploring vegetable ferments to create savoury condiments that pair well with meats or enhance vegetable-based dishes. The beauty of fermentation lies in its adaptability, allowing chefs to customize the process to suit the flavours they want to achieve.

Drying is another technique that has found a place in modern kitchens, particularly for enhancing the texture and flavour of ingredients. Dehydrated fruits and vegetables are used as garnishes, snacks, and flavour boosters, adding concentrated sweetness or umami to dishes. Powdered ingredients, such as dried mushrooms or tomatoes, are often used to intensify flavour in sauces, soups, and stews. For instance, dried mushroom powder adds depth to vegetarian dishes, providing a rich, earthy flavour without the need for meat. Chefs can also create "house-dried" products to showcase the unique flavours of local ingredients, giving their dishes a distinctive sense of place. Beyond flavour, drying techniques enable chefs to reduce food waste, as overripe fruits or surplus vegetables can be dried and preserved, contributing to sustainable kitchen practices.

Salting, a technique that might seem simple, continues to inspire chefs, particularly in the art of curing meats and seafood. Charcuterie boards, featuring salted and cured meats like prosciutto, salami, and lox, have become popular, offering diners a taste of traditional preservation methods. Cured meats not only have a longer shelf life but also showcase complex flavours achieved through salt's transformative power. In addition to charcuterie, salting is used to cure fish, creating delicacies such as gravlax (salt-cured salmon) or salted cod, which can be prepared in various ways. Salt curing allows chefs to explore the nuances of flavour and texture, adding depth to their menus while paying homage to ancient preservation practices.

Modern chefs are also rediscovering smoking, often combined with salting and drying, to add unique flavours and preserve ingredients. Smoked meats, fish, and even vegetables add depth and aroma to dishes, enhancing both taste and presentation. Some chefs are experimenting with cold-smoking techniques to preserve the natural texture of foods while adding a subtle smoky flavour, while others are using wood from specific trees—like apple or cherry wood—to impart distinctive aromas. Smoking also allows chefs to create unique "smoked" versions of foods not traditionally associated with the technique, such as butter or even cocktails, adding innovative twists to traditional meals.

Beyond flavour, these traditional preservation techniques contribute to sustainable culinary practices by reducing food waste and extending the shelf life of ingredients. By preserving foods, chefs can make use of seasonal surpluses, prevent spoilage, and create ingredients that add value to their kitchens over time. For example, during peak harvest seasons, chefs might ferment or dry excess produce to create ingredients that can be used year-round. This approach aligns with the growing movement toward sustainability in the culinary world, as chefs seek ways to minimize waste and celebrate seasonal produce.

The revival of traditional preservation techniques in modern culinary practices reflects both a desire to reconnect with food heritage and a commitment to sustainable, flavourful cooking. By incorporating methods like fermentation, drying, and salting, chefs can expand their culinary repertoire, creating dishes that are not only rich in flavour but also rooted in history and cultural tradition. These methods remind chefs of the ingenuity of past societies and offer a means of bridging the past with the present, honouring the timeless connection between food preservation and culinary creativity.

CHAPTER 7: THE INFLUENCE OF TRADE ON CULINARY PRACTICES

The Silk Road, Spice Routes, And Their Role In Global Culinary Exchanges

The development of ancient trade routes such as the Silk Road and spice routes fundamentally altered the culinary landscapes of civilizations across the world. These routes, stretching across continents and connecting distant cultures, enabled the exchange of spices, grains, fruits, and culinary techniques, all of which have left a lasting imprint on global cuisines. Beyond the transport of goods, these trade routes served as channels of cultural diffusion, linking regions through shared tastes and food practices.

The Silk Road was one of the earliest and most extensive

networks of overland and maritime routes. Beginning around the 2nd century BCE and extending until the 14th century, this route connected China with the Middle East, North Africa, and Europe. While the Silk Road was named after the Chinese silk that was highly prized and widely traded, it carried far more than textiles. Foods, spices, and techniques flowed alongside silk, fostering a culinary exchange that influenced diets from Asia to Europe. Tea, one of China's treasured exports, became popular in Central Asia and beyond, eventually making its way to the courts of the Middle East and Europe. Similarly, rice, another staple that travelled along the Silk Road, became a dietary cornerstone in regions far from its origins, influencing culinary practices in the Middle East and eventually Europe.

In addition to ingredients, the Silk Road facilitated the exchange of cooking techniques. For instance, noodle-making, a staple of Chinese cuisine, was introduced to Persia through these routes. This technique gradually spread westward, influencing pasta-making traditions in Italy and other parts of Europe. The early influences of these exchanges are still visible today in dishes like Italian spaghetti, which has roots in early Chinese noodle-making practices, albeit adapted over centuries. Even bread-making techniques saw adaptation and diffusion along the Silk Road, with leavened and unleavened breads moving between the Middle East, Central Asia, and East Asia.

The spice routes, often referred to as the maritime routes, stretched across the Indian Ocean and connected India, the Arabian Peninsula, and East

Africa with Europe. Spices like black pepper, cinnamon, cloves, nutmeg, and cardamom were the primary drivers of these routes, prized in Europe for their ability to enhance flavours and preserve foods. These routes sparked some of the earliest forms of long-distance trade and were instrumental in the culinary evolution of many cultures. The high demand for spices in Europe led to the establishment of trading outposts, colonies, and, eventually, entire trade empires controlled by European powers. The Portuguese, for instance, monopolized the spice trade in the early 16th century by controlling the trade routes between India and Europe, bringing spices such as black pepper and cinnamon to European tables.

The presence of spices in medieval Europe became a symbol of wealth and sophistication. Noble households would feature exotic spices in their dishes, using them not only to add flavour but to demonstrate their social status. Cloves, nutmeg, and pepper became integral to European cuisine, particularly in sauces, stews, and pastries, where their pungent flavours transformed otherwise simple dishes. The influence of these spice routes extended into Middle Eastern cuisine as well. Spices became staples in Arab and Persian cooking, contributing to richly spiced dishes such as Persian stews, Middle Eastern meat pies, and Indian curries. For centuries, the spice trade brought about a continuous cultural exchange that enriched culinary traditions across the world, giving rise to iconic spice blends such as garam masala in India, baharat in the Middle East, and ras el hanout in North Africa.

Notable Ingredients That Transformed Global Cuisines (e.g., Spices, Cocoa, Coffee)

The impact of the global trade routes can be most vividly seen in the way certain ingredients reshaped the culinary landscapes of the regions they reached. Among these transformative ingredients, spices, cocoa, and coffee stand out for their enduring influence on food and culture.

Spices were among the earliest and most sought-after trade goods, their exotic flavours, aromas, and preservative qualities making them invaluable in the culinary practices of many civilizations. Black pepper, often referred to as "black gold," was one of the first spices to travel from its native India to Europe, where it became a staple seasoning among the elite. This tiny, pungent seed not only added heat and depth to dishes but also played a role in preserving food, particularly meats. Over time, black pepper became a cornerstone of European cuisine, influencing everything from soups and sauces to roasted meats. Similarly, cinnamon, native to Sri Lanka, transformed desserts and stews alike, adding a sweet, warming quality that quickly became popular across Middle Eastern and European cuisines.

Nutmeg and cloves, both native to the Maluku Islands (the "Spice Islands") in Indonesia, were highly prized for their unique flavours and medicinal properties. Nutmeg, with its warm and slightly sweet flavour, was used in both sweet and savoury dishes, particularly in the European and Middle Eastern kitchens, while cloves added an aromatic warmth to everything from meats to desserts. The high demand for these spices in Europe drove the establishment of Dutch and Portuguese colonies in Southeast Asia, sparking competition between European powers to control these valuable commodities. The impact of these spices on European cuisine was profound, leading to the development of holiday treats such as gingerbread and mulled wine, which used cloves and nutmeg to create comforting, aromatic flavours.

Cocoa, indigenous to the Americas, had a dramatic effect on global cuisine when it was introduced to Europe by Spanish explorers in the 16th century. In its original form, cocoa was consumed by the Mayans and Aztecs as a bitter, spiced drink, often mixed with chili peppers and used in ceremonial rituals. When the Spanish encountered cocoa, they adapted it by adding sugar and milk, transforming it into the sweet, creamy chocolate beverage that would become a European favourite. Over time, European chocolatiers developed new methods for processing cocoa, turning it into solid chocolate bars and confections, which became symbols of luxury and sophistication. Today, chocolate is used in a wide variety of dishes, from desserts and pastries to savoury sauces like Mexican mole, a rich, complex sauce that combines cocoa with spices and chilies, highlighting the fusion of Indigenous and Spanish culinary traditions.

Coffee, native to the highlands of Ethiopia, is another example of a transformative trade ingredient. It spread from Ethiopia to Yemen, where coffee cultivation began, and from there, it travelled throughout the

Middle East and into Europe by the 17th century. Coffeehouses soon became a fixture in cities such as Mecca, Cairo, and Istanbul, where they served as gathering places for intellectuals, artists, and merchants. Known as "schools of the wise," these coffeehouses became hubs for social and cultural exchange, where people gathered to discuss philosophy, politics, and art. As coffee spread to Europe, it played a similar role, with coffeehouses in cities like Paris, London, and Vienna becoming important centres of social and intellectual life. Today, coffee remains a globally beloved beverage, with each region developing its unique methods of brewing and serving it, from Italian espresso to Turkish coffee and American pour-overs.

These ingredients—spices, cocoa, and coffee—revolutionized the culinary landscapes of the regions they entered, introducing new flavours, textures, and rituals that have become deeply embedded in global cuisine. Their legacy illustrates the power of culinary exchange to transform not only diets but also social and cultural practices, shaping the way people experience and enjoy food around the world.

How Chefs Today Can Draw Inspiration from the Rich History of Food Trade

The legacy of historical trade routes offers a wealth of inspiration for modern chefs, who can draw on this history to create dishes that honour the fusion of flavours and ingredients that have defined global cuisine. By understanding the journey of ingredients and the ways they have been adapted and transformed by different cultures, chefs can infuse their menus with depth, authenticity, and cultural appreciation.

One way chefs can honour the history of food trade is by experimenting with spice blends and flavours from various regions. Creating spice blends inspired by the spice routes can add unique dimensions to dishes and evoke the spirit of global culinary exchange. For example, a chef could combine spices traditionally associated with Indian and Middle Eastern cooking, such as turmeric, coriander, cumin, and sumac, to create a new blend that reflects the cross-cultural fusion seen along historical trade routes. This approach allows chefs to bring complexity and diversity to their menus, enhancing dishes with flavours that speak to a broader cultural context.

Chefs can also experiment with fusion dishes that combine ingredients and techniques from different culinary traditions. One of the most celebrated examples of fusion cuisine is the Nikkei cuisine of Peru, which blends Japanese and Peruvian ingredients and techniques. This cuisine emerged in the early 20th century when Japanese immigrants brought their culinary traditions to Peru, creating dishes like ceviche with soy sauce and ginger. Similarly, chefs today can create dishes that blend elements from different cultures to pay homage to the history of food trade. For example, a chef might prepare pasta with an Asian-inspired sauce that incorporates miso, sesame oil, and ginger, blending Italian and Japanese influences in a dish that reflects the spirit of culinary exchange.

Reimagining traditional dishes using ingredients from

other cultures is another way chefs can draw inspiration from trade history. For instance, classic European desserts like panna cotta or crème brûlée can be infused with spices like cardamom or saffron, adding an exotic twist while preserving the integrity of the original dish. By incorporating foreign ingredients in traditional recipes, chefs create dishes that celebrate global flavours and demonstrate the adaptability of culinary traditions. This approach acknowledges the influence of trade while adding a modern and personal touch to traditional flavours.

Educating diners about the origins and cultural significance of ingredients enhances the dining experience, allowing guests to appreciate the history behind the flavours on their plates. By sharing the stories of ingredients and how they travelled from one region to another, chefs create an immersive experience that connects diners to the broader history of food. For instance, a chef serving mole, the iconic Mexican sauce with chocolate and chilies, might explain its roots in Aztec cuisine and the influence of Spanish colonialism, helping diners understand the layers of history in each bite. This educational aspect enriches the dining experience, transforming it into a journey through time and culture.

Promoting sustainability and supporting local farmers reflects the spirit of historical trade practices that respected the land and its resources. In ancient trade systems, communities relied on locally sourced ingredients and engaged in mutually beneficial exchanges with neighbouring regions. Modern chefs can honour this legacy by sourcing ingredients from local farms and practicing sustainable cooking. For example, chefs can feature regional herbs, fruits, and grains in their menus, highlighting local flavours while reducing the environmental impact of imported goods. By supporting local producers, chefs not only contribute to their communities but also echo the values of early trade practices that emphasized respect for the earth's bounty.

Collaborating with chefs from different cultural backgrounds offers another opportunity for chefs to engage with the legacy of trade routes. By working alongside chefs who bring expertise in other culinary traditions, chefs can learn new techniques, explore unfamiliar ingredients, and create fusion dishes that are both respectful and innovative. These collaborations embody the spirit of cultural exchange that defined historical trade routes, allowing chefs to expand their culinary horizons and introduce diners to new flavours and experiences.

The history of food trade provides modern chefs with a rich tapestry of flavours, techniques, and stories to draw from. By experimenting with global spice blends, creating fusion dishes, reimagining traditional recipes, educating diners, supporting sustainable practices, and collaborating with other chefs, they can celebrate the enduring legacy of culinary exchange. This approach not only honours the history of food trade but also inspires chefs to continue the journey of culinary innovation, bridging cultures and creating connections through the universal language of food.

CHAPTER 8: CULINARY TECHNIQUES FROM EAST ASIA

Overview Of Culinary Techniques And Ingredients In China, Japan, And Korea

East Asia is home to some of the world's most revered culinary traditions, each shaped by unique philosophies, ingredients, and techniques. The cuisines of China, Japan, and Korea are particularly renowned for their distinct flavours, meticulous preparation methods, and symbolic presentations. Each of these cultures brings its own perspective to the art of cooking, with specific techniques and ingredients that define their culinary identity.

In China, one of the oldest and most influential culinary cultures, the approach to food preparation is guided by principles of balance, harmony, and contrast. Chinese cuisine places a strong emphasis on texture, colour, and flavour, often combining sweet, salty, sour, and spicy tastes in a single meal to create a well-rounded sensory experience. Traditional Chinese cooking relies on a variety of techniques, including stir-frying, steaming, braising, and deep-frying, each used to bring out the best qualities in ingredients. Rice and wheat are staple grains in Chinese cuisine, along with a wide range of vegetables, meats, and tofu. The importance of soy-based products, such as soy sauce, tofu, and fermented black beans, is particularly noteworthy, as they add umami and depth to dishes, enhancing the complexity of flavours. Aromatics like ginger, garlic, and green onions are frequently used, creating a foundation of flavours that is both fragrant and flavourful.

Japan, in contrast, is known for its minimalistic approach to cooking, where the goal is to enhance the natural flavours of fresh, high-quality ingredients rather than mask them. Japanese cuisine emphasizes seasonality (known as "shun") and harmony, following the principles of "washoku," a culinary philosophy that promotes balance and simplicity. Japanese techniques include steaming, grilling, simmering, and precise knife work, with a focus on presenting food in a way that reflects the aesthetics of nature. Rice is central to the Japanese diet, accompanied by seafood, vegetables, and seasonal ingredients. Sushi, sashimi, and tempura are some of the iconic dishes that embody Japanese culinary principles. Fermented products like miso, soy sauce, and mirin add complexity to Japanese cuisine, creating subtle layers of umami that enhance the natural taste of ingredients.

Korean cuisine is characterized by bold, robust flavours, with an emphasis on fermented ingredients that provide a distinctive tang and depth. The art of fermentation, particularly with dishes like kimchi and gochujang (fermented chili paste), is central to Korean cooking, imparting both flavour and probiotic benefits. Korean cuisine is also known for its extensive use of garlic, sesame oil, chili, and soy sauce, which together create a rich, spicy, and savoury profile. Korean meals are typically served with a variety of side dishes (banchan), including pickled and fermented vegetables, which complement the main dishes and provide a balanced array of flavours and textures. Grilling and braising are also key techniques, with dishes like bulgogi (grilled marinated beef) and bibimbap (a mixed rice dish) showcasing Korea's distinctive approach to seasoning and ingredient preparation.

Each of these culinary traditions reflects the region's unique ingredients, agricultural practices, and cultural values. Understanding these culinary foundations allows chefs to appreciate the intricacies of East Asian cooking, from the balance and harmony of Chinese flavours to the seasonal mindfulness of Japanese dishes and the bold, fermented flavours of Korean cuisine.

Key Cultural Dishes and Techniques Such as Steaming, Fermenting, and Precision Knife Skills

The culinary techniques of East Asia have been refined over centuries, each technique honed to bring out the best qualities of the ingredients and create a harmonious dining experience. Some of the most notable techniques include steaming, fermenting, and precision knife skills, each of which holds cultural significance and reflects the philosophical underpinnings of the respective cuisines.

Steaming is a foundational technique in Chinese cooking, used to preserve the natural flavours and nutritional value of ingredients. In China, steaming is often used to prepare delicate dishes like dim sum, fish, and vegetables. Dim sum, for instance, includes an array of steamed dumplings, buns, and rolls, each meticulously crafted and often served in bamboo steamers, which impart a subtle aroma to the food. Steaming is also common in Japanese and Korean cuisines, where it is used to cook rice and prepare delicate dishes like chawanmushi (Japanese savoury egg custard) and steamed fish. The steaming technique requires careful temperature control and timing to ensure the ingredients are cooked evenly without losing moisture. For East Asian chefs, steaming embodies the value of respect for natural flavours, as it allows the ingredients to shine without excessive seasoning or intervention.

Fermentation is an ancient technique practiced widely in both Korea and Japan, where it has become an essential part of the food culture. In Korea, the process of fermenting vegetables, especially cabbage and radishes, has given rise to the iconic dish kimchi, which is made by layering vegetables with chili paste, garlic, ginger, and other seasonings, then allowing it to ferment. Kimchi is not only a dietary staple but also a symbol of Korean heritage, with each family often having its own recipe. Fermentation is also used to make gochujang (spicy fermented chili paste) and doenjang (fermented soybean paste), which form the backbone of Korean seasoning. In Japan, the art of fermentation is seen in products like miso and soy sauce, both of which add depth and umami to Japanese dishes. Miso, made from fermented soybeans and rice or barley, is used in soups, marinades, and sauces, enhancing both flavour and nutritional value. Fermentation in East Asia goes beyond preservation —it adds complexity and creates a unique profile of flavours, textures, and aromas that are integral to these cuisines.

Precision knife skills are another hallmark of East Asian culinary practices, particularly in Japanese cuisine, where knife work is elevated to an art form. In Japan, different types of knives are used for specific tasks, each crafted for a particular ingredient or cutting style. For example, the yanagiba, a long, slender knife, is used for slicing sashimi, while the deba, a heavy-duty knife, is designed for cutting fish bones. The precise cutting techniques ensure that each slice is clean and uniform, allowing the natural flavours and textures of the ingredients to be enjoyed without interference. This meticulous approach to knife work reflects the Japanese philosophy of respect and mindfulness, where the act of cutting becomes a meditative practice. In Chinese cuisine, knife skills are also essential, particularly in stir-frying, where ingredients are cut to uniform sizes to ensure even cooking. The Chinese cleaver is a versatile tool used for chopping, slicing, and dicing, and its wide, flat surface is often used to transfer ingredients to the wok.

These techniques—steaming, fermenting, and precision knife work—are not merely methods of cooking but expressions of cultural values and culinary philosophies. For chefs interested in East Asian techniques, mastering these methods provides insight into the art of restraint, respect for ingredients, and the pursuit of harmony in cooking.

Applications for Chefs Interested in East Asian Culinary Techniques

For chefs looking to incorporate East Asian techniques into their culinary practice, the methods of steaming, fermenting, and precision knife skills offer valuable tools for creating dishes that are both flavourful and culturally resonant. By understanding the philosophical principles underlying these techniques, chefs can enhance their approach to cooking, drawing inspiration from the rich traditions of East Asia while adapting them to modern cuisine.

Steaming offers chefs a way to cook ingredients delicately, preserving their natural flavours, textures, and nutritional value. Modern chefs can use steaming to prepare a wide range of dishes, from seafood and vegetables to dumplings and desserts. Steaming is particularly well-suited for health-conscious cooking, as it requires little to no oil and maintains the freshness of the ingredients. Chefs can experiment with steaming different types of vegetables, proteins, and even grains, using bamboo steamers for an added layer of aroma and presentation. For a fusion approach, chefs might incorporate flavours like lemongrass, ginger, or miso into their steamed dishes, blending East Asian techniques with local ingredients. Steaming also allows chefs to create visually appealing presentations, as the natural colours of the ingredients are retained, making the dishes look vibrant and fresh.

Fermentation provides chefs with an opportunity to introduce depth and complexity to their dishes. By experimenting with fermentation, chefs can create house-made condiments like kimchi, miso, or pickled vegetables, adding unique flavours and textures to their menus. For instance, chefs might use kimchi as a garnish or condiment, adding a tangy, spicy element to a dish, or incorporate miso into marinades, sauces, and dressings to add umami. For those new to fermentation, starting with simple pickling techniques can be an accessible way to explore the world of fermented foods. Pickling cabbage, radishes, or cucumbers in a brine seasoned with rice vinegar, salt, and chili offers a simple yet effective way to add an East Asian-inspired touch to a dish. Fermentation also aligns with sustainable cooking practices, as it allows chefs to preserve seasonal produce and reduce waste.

Precision knife skills are a valuable asset for any chef, and East Asian knife techniques provide a framework for improving one's approach to cutting, slicing, and presenting food. By practicing Japanese knife skills, chefs can elevate their presentation, making dishes visually stunning and precise.

Learning to use different types of knives, such as a yanagiba for slicing raw fish or a usuba for cutting vegetables, adds authenticity to East Asian-inspired dishes and enhances the dining experience. These knife skills can be applied beyond traditional East Asian dishes; for example, chefs can use the slicing techniques from sashimi preparation to create delicate cuts of fruits or vegetables, adding an artistic touch to salads, appetizers, and garnishes. The attention to detail required in Japanese knife skills teaches chefs to work mindfully, focusing on the quality and presentation of each ingredient.

For chefs interested in deeper engagement with East Asian cuisine, studying regional variations within China, Japan, and Korea offers additional insight. In China alone, the cuisines of Sichuan, Cantonese, and Shandong regions each have distinct flavour profiles and techniques. Sichuan cuisine is known for its bold, spicy flavours and the use of Sichuan peppercorn, which creates a unique numbing sensation. Cantonese cuisine, by contrast, emphasizes light, fresh flavours and delicate cooking techniques like steaming and poaching. Exploring these regional styles allows chefs to appreciate the diversity within Chinese cuisine and draw inspiration from its vast culinary heritage.

In Japan, regional differences in cooking reflect the country's geography and climate. The cuisine of Hokkaido, for example, is known for its seafood and dairy products, while the Kansai region specializes in dishes like okonomiyaki (savoury pancakes) and takoyaki (octopus balls). Korean cuisine also varies by region, with the southern province of Jeolla known for its abundant use of banchan, and the northern region famous for its milder flavours and reliance on soy-based seasonings. For chefs, studying these regional distinctions within East Asian cuisines provides a broader palette of flavours and techniques, enriching their ability to create diverse and authentic dishes.

East Asian culinary techniques offer chefs a range of methods to enhance their cooking, from the subtle art of steaming to the bold flavours of fermentation and the precision of knife skills. By embracing these techniques, chefs can bring the philosophy of East Asian cuisine into their own kitchens, creating dishes that honour both the simplicity and complexity of these traditions. For chefs inspired by the culinary heritage of East Asia, these techniques provide a pathway to a richer, more mindful approach to cooking, blending tradition with innovation in a way that respects the artistry of each ingredient.

CHAPTER 9: INDIGENOUS CUISINES OF THE AMERICAS

Overview Of Indigenous Foods And Their Impact On Modern Cuisines

The Indigenous peoples of the Americas cultivated a wide variety of foods that have shaped the diets and culinary traditions of societies around the world. Long before European contact, Indigenous communities across the Americas developed advanced agricultural systems and culinary practices that utilized their native flora and fauna. Corn, beans, squash, potatoes, tomatoes, and cacao are just a few examples of foods that originated in the Americas and have since become staples in global cuisine. Indigenous foods have had a lasting impact, both nutritionally and culturally, offering flavours, ingredients, and techniques that are still celebrated in contemporary kitchens.

One of the most significant Indigenous contributions

to modern cuisine is the "Three Sisters" agricultural system, a method of planting corn, beans, and squash together to maximize their growth and nutrient yield. This technique, widely practiced by Indigenous tribes across North America, involved planting corn as a support structure for climbing beans, while the broad leaves of squash covered the soil to retain moisture and reduce weeds. The Three Sisters not only supported sustainable agriculture but also provided a complete nutritional profile, with corn providing carbohydrates, beans adding protein, and squash offering vitamins and minerals. This trio of crops has remained central to Indigenous food traditions and has influenced farming practices worldwide.

In addition to staple crops, Indigenous communities used their vast knowledge of local plants and ecosystems to create a wide variety of flavours and textures in their diets. They cultivated fruits such as strawberries, blueberries, and avocados, as well as herbs like sage and cilantro, many of which are now integral to modern culinary practices. They also harvested wild plants, nuts, and seeds, relying on biodiversity to create a diet that was resilient and varied. The knowledge and respect Indigenous peoples held for the land are reflected in their culinary practices, which emphasized seasonal eating, waste reduction, and sustainable harvesting.

The Indigenous culinary heritage of the Americas also includes sophisticated cooking techniques, such as roasting, steaming, and fermenting. Many Indigenous cultures used clay pots for cooking, an early form of slow-cooking that infused food with earthy flavours.

They also developed methods of preserving food, such as drying and smoking, to extend the shelf life of meats, fruits, and grains. Indigenous methods like smoking are still widely used today in barbecuing and other cooking styles, providing modern chefs with valuable insights into preservation and flavour development. The foods and techniques developed by Indigenous communities laid the foundation for many culinary practices that are now common in contemporary kitchens, demonstrating the lasting influence of Indigenous cuisines on global food culture.

Case Study of Mesoamerican, Andean, and Native American Food Traditions

The Americas encompass a wide range of Indigenous cultures, each with unique culinary traditions. Examining the food practices of Mesoamerican, Andean, and Native American cultures provides insight into the diversity and depth of Indigenous cuisines, each rooted in its specific landscape and environment.

Mesoamerican cuisine—which includes the culinary traditions of ancient civilizations like the Maya and Aztec—has had a profound influence on modern cooking, especially in Mexico and Central America. Corn, or maize, was the staple crop of Mesoamerican civilizations, regarded as a sacred gift from the gods. The process of nixtamalization, in which corn is soaked and cooked in an alkaline solution, was developed to improve the nutritional value of maize. This method allowed the release of essential nutrients, making corn a more complete food source. Tortillas, tamales, and atole (a traditional corn-based drink) are all examples of foods created through nixtamalization. Additionally, cacao was highly valued by the Maya and Aztecs, who used it as currency and in ceremonial rituals. The traditional Mesoamerican chocolate drink, made from cacao beans, chili, and water, has evolved into the hot chocolate and chocolate-based desserts we enjoy today. These foods showcase the innovation of Mesoamerican culinary practices and their ongoing cultural significance.

Andean cuisine reflects the unique environment of the Andes Mountains, where Indigenous people developed a diet that included potatoes, quinoa, and guinea pig (cuy), which remains a traditional source of protein. Potatoes, which originated in the Andean region, were cultivated in hundreds of varieties, each suited to specific microclimates in the mountains. Indigenous Andeans developed freeze-drying techniques to preserve potatoes, creating chuño, a dehydrated potato that could be stored for years. Quinoa, known as the "mother grain," was another staple crop, prized for its high protein content and adaptability to high-altitude conditions. The Andean peoples also grew ají peppers, which added heat and flavour to their dishes. Today, quinoa and potatoes are enjoyed worldwide, but their origins in the Andes highlight the resourcefulness of Indigenous communities who adapted their diets to a challenging environment.

Native American cuisine in North America encompasses a wide range of traditions, from the seafood-rich diets of coastal tribes to the bison-based diet of the Plains tribes. Indigenous tribes of the northeastern woodlands practiced the Three Sisters farming method and often incorporated wild game, berries, and maple syrup into their diets. Coastal tribes, such as the Pacific Northwest Coast peoples, relied on the abundant fish and shellfish available in their waters, using techniques like smoking and drying to preserve their catch. The Plains tribes, including the Lakota and Cheyenne, hunted bison and developed methods for processing and preserving the meat, often drying it into pemmican, a high-protein food made by combining dried bison meat with fat and berries. These culinary practices are tied deeply to the land and seasons, reflecting each community's relationship with their environment. Many of these foods—such as smoked salmon, pemmican, and maple syrup—continue to be celebrated and enjoyed today, both within Indigenous communities and beyond.

These case studies illustrate the diversity and adaptability of Indigenous food traditions across the Americas. Each culture developed its unique approach to agriculture, ingredient selection, and food preparation, all of which were intimately connected to the landscape. For chefs, these traditions offer a source of inspiration and a reminder of the deep respect Indigenous peoples have for the environment.

How Chefs Can Incorporate Indigenous Ingredients and Practices Respectfully

As the global culinary community increasingly recognizes the value of Indigenous foods and practices, chefs have an opportunity—and a responsibility—to incorporate these traditions with cultural sensitivity and respect. Integrating Indigenous ingredients and methods into contemporary cuisine requires not only technical skill but also an understanding of the historical and cultural significance of these foods. By approaching Indigenous culinary traditions thoughtfully, chefs can celebrate these rich heritages while honouring the communities from which they originate.

One of the most important principles for chefs interested in Indigenous cuisine is to source ingredients responsibly and respectfully. Many Indigenous foods, such as wild rice, bison, and heirloom corn varieties, have cultural and ecological importance to Native communities. Sourcing these ingredients from Native-owned suppliers or community-based initiatives helps support Indigenous food sovereignty and respects the rights of Indigenous peoples to steward their traditional foods. For example, chefs who wish to use wild rice might source it from Native American tribes in the Great Lakes region, where wild rice has been cultivated and harvested for centuries as a sacred food. By sourcing ingredients in ways that honour the Indigenous communities who cultivate them, chefs can contribute to the preservation of these food traditions.

Another way chefs can incorporate Indigenous practices respectfully is by studying the traditional preparation methods and understanding their cultural significance. Techniques like nixtamalization, for instance, are not only practical but are also deeply rooted in Mesoamerican culture and spirituality. Chefs can experiment with nixtamalization to create authentic tortillas or tamales, appreciating the labour-intensive process that has been perfected over centuries. Similarly, chefs can explore methods of cooking with clay pots, an ancient technique used by Indigenous peoples to infuse dishes with a unique flavour and earthiness. Understanding the cultural context behind these methods is crucial, as it allows chefs to approach these techniques with an awareness of their origins and importance.

Collaborating with Indigenous chefs and community members is another meaningful way to incorporate Indigenous practices respectfully. By working with individuals who are part of these communities, chefs can gain a more authentic understanding of Indigenous culinary traditions and ensure that their interpretations are accurate and respectful. Collaboration can take many forms, from hosting guest Indigenous chefs for pop-up events to attending workshops led by Indigenous food experts. These interactions provide valuable insights and foster a sense of connection, bridging the gap between traditional and contemporary culinary practices. For example, a chef might invite an Indigenous expert to discuss the cultural significance of corn varieties, offering diners an educational experience that enriches their understanding of Indigenous food heritage.

When creating dishes inspired by Indigenous cuisine, it is also important for chefs to acknowledge the origins of the ingredients and methods used. By including information on menus or engaging in conversations with diners, chefs can educate their clientele about the history and significance of Indigenous foods. For instance, a menu item might explain that the corn used in a dish is a heritage variety grown by Indigenous farmers, or that a sauce was made using a traditional Andean chili. This approach not only celebrates Indigenous heritage but also helps combat the erasure of Indigenous contributions to global food culture.

Chefs can also honour Indigenous culinary traditions by adopting sustainable practices that reflect the ecological values of many Indigenous cultures. Traditional Indigenous food systems are often based on principles of sustainability, including minimal waste, seasonal eating, and respect for the land. By embracing these values, chefs can create menus that are environmentally conscious and aligned with Indigenous philosophies. Using seasonal ingredients, reducing food waste, and incorporating foraged foods can all be part of a respectful approach to Indigenous-inspired cuisine. This alignment with Indigenous values goes beyond technique, embodying a holistic approach to cooking that respects the earth and its resources.

Finally, chefs should approach Indigenous cuisine with humility and an openness to learning. Indigenous food traditions are deeply rooted in history, spirituality, and community, and it is essential for chefs to recognize the significance of these connections. By acknowledging the limitations of their understanding and seeking guidance from Indigenous voices, chefs can create a respectful and authentic representation of Indigenous cuisine. This humility fosters a more meaningful connection to the food and ensures that Indigenous contributions are celebrated rather than appropriated.

Indigenous cuisines of the Americas offer a rich array of ingredients, techniques, and philosophies that can inspire chefs to create dishes with depth and purpose. By sourcing ingredients responsibly, respecting traditional methods, collaborating with Indigenous communities, educating diners, and embracing sustainability, chefs can incorporate Indigenous practices in a way that honours the cultural and ecological heritage of these foods. Through thoughtful engagement, chefs have the opportunity to celebrate Indigenous culinary traditions and contribute to the broader recognition and appreciation of these invaluable food cultures.

CHAPTER 10: AFRICAN CULINARY TRADITIONS

Exploration Of African Culinary Techniques And Staple Ingredients

African cuisine, diverse and deeply rooted in regional traditions, is characterized by a wealth of ingredients and techniques that reflect the continent's unique landscapes, climates, and cultural diversity. African culinary practices have evolved over millennia, shaped by agricultural practices, indigenous knowledge, and cultural exchanges through trade and migration. While each region of Africa has its own distinctive culinary practices, some common techniques and staple ingredients bind them together, creating a culinary heritage that is rich, flavourful, and complex.

Traditional cooking techniques are fundamental to African cuisine. Roasting, boiling, steaming, frying, and stewing are common across the continent, often performed in ways that enhance the natural flavours of ingredients and preserve their nutrients. One prominent technique is grilling or roasting over open fire, which gives meats and vegetables a distinct smoky flavour. In West Africa, meats are often marinated in spicy sauces before being roasted or grilled, a technique that imbues dishes with bold, layered flavours. Another technique is slow-cooking stews in clay pots, a method that allows ingredients to meld over a low flame, creating rich, hearty dishes. Many African stews are built around tomatoes, onions, and peppers, a trio that serves as the flavour base for many dishes across the continent, particularly in West and Central Africa.

Fermentation is another significant technique in African culinary traditions, particularly in West and East Africa, where fermented grains and dairy are common. Fermentation is used to make foods like ogiri (a fermented seed condiment) in West Africa and inji in Ethiopia, a tangy fermented flatbread made from teff flour. Fermented foods have long been valued for their nutritional benefits and their ability to enhance flavours, creating complex, umami-rich dishes that are deeply satisfying. The fermentation of grains also allows for long-term storage, which is crucial in regions where food scarcity and preservation remain important.

Africa's staple ingredients are as diverse as its landscapes, with each region relying on a different set of crops based on the local climate and agricultural practices. Cassava, yams, maize, millet, and sorghum are some of the continent's most important staple crops, providing sustenance for millions and forming the foundation of many African dishes. Cassava, for instance, is a root vegetable that is widely consumed in Central and West Africa. It can be boiled, roasted, or ground into flour to make fufu, a starchy dough that serves as an accompaniment to stews and soups. Yams and sweet potatoes are also widely used, adding substance and sweetness to dishes. Millet and sorghum, resilient grains that thrive in arid conditions, are essential in East and West African diets, where they are used to make porridges, flatbreads, and even fermented beverages.

Legumes, such as cowpeas, black-eyed peas, and peanuts, are another staple in African diets, providing protein and texture to dishes. Peanuts, in particular, are a central ingredient in many West African stews, adding creaminess and depth to dishes like maafe (peanut stew). Peanuts are also ground into pastes and sauces, used to flavour everything from vegetables to meats. Additionally, Africa is home to a variety of indigenous fruits and vegetables, including okra, bitter leaf, baobab fruit, and plantains, each adding unique flavours and nutritional benefits to traditional dishes. Okra, for instance, is often used to thicken soups and stews, while plantains—cooked in various forms from fried to boiled—are an essential part of many African diets.

Together, these techniques and ingredients form the backbone of African cuisine, showcasing the resourcefulness, adaptability, and depth of African culinary traditions. For chefs, understanding these fundamental elements provides a gateway to exploring the rich flavours and textures that African cuisine has to offer.

Regional Flavours and Dishes from West, East, and North Africa

Africa's regional cuisines are as varied as its geography, with each region offering unique dishes and flavour profiles that reflect local ingredients, traditions, and influences. Examining the cuisines of West, East, and North Africa provides insight into the diversity and vibrancy of African food traditions.

West African cuisine is known for its bold, spicy flavours, often created through the use of hot peppers, ginger, and various herbs. A popular dish across the region is jollof rice, a one-pot rice dish cooked with tomatoes, onions, and spices, often served with grilled or stewed meat. Jollof rice has become an iconic dish in countries like Nigeria, Ghana, and Senegal, with each region adding its unique twist to the recipe. Another signature dish is egusi soup, made from ground melon seeds and typically served with fufu or pounded yam. The soup is thick and hearty, featuring leafy greens, meat, and fish, and seasoned with a blend of spices that give it a rich, nutty flavour. Suya, a popular street food in Nigeria, is another highlight of West African cuisine. It consists of skewered meat seasoned with a peanut-spice rub, then grilled over an open flame, creating a smoky, spicy snack that is both flavourful and satisfying.

East African cuisine is marked by its use of spices and grains, with influences from Arabic, Indian, and Persian culinary traditions. This region is known for dishes like injera, a fermented flatbread made from teff, which is a staple in Ethiopian and Eritrean cuisine. Injera is often served with various stews known as wot, made with lentils, chickpeas, vegetables, and meats, seasoned with a distinctive spice blend called berbere. In coastal regions, such as Kenya and Tanzania, seafood plays a prominent role, often seasoned with spices like cloves, cardamom, and cinnamon, reflecting the influence of the Indian Ocean spice trade. Ugali, a type of maize porridge, is a staple food across East Africa, served with vegetables, meat stews, or fish. The diversity of East African cuisine reflects the region's geographic location as a hub for cultural exchange, where spices, ingredients, and techniques from different cultures have melded to create a unique culinary landscape.

North African cuisine is renowned for its use of aromatic spices and herbs, as well as for its richly flavoured stews and tagines. Influenced by Mediterranean, Berber, and Arab cultures, North African food is both fragrant and visually appealing. One of the region's signature dishes is couscous, a semolina-based grain that is steamed and served with stewed meats and vegetables. In Morocco, tagine dishes —named after the clay pot they are cooked in— are popular, combining meats, vegetables, dried fruits, and spices like saffron, cumin, and cinnamon. Tagines are slow-cooked, allowing the flavours to meld and intensify, creating a rich, complex dish. Harira, a tomato-based soup with lentils and chickpeas, is a staple in Moroccan and Algerian cuisine, often enjoyed during the holy month of Ramadan. In Egypt, ful medames, a dish made from fava beans stewed with garlic, lemon, and olive oil, is a popular breakfast food that dates back to ancient times.

Each of these regions showcases unique flavours, techniques, and ingredients that highlight Africa's culinary diversity. West Africa's bold, spicy dishes, East Africa's fusion of indigenous and imported flavours, and North Africa's aromatic, spice-laden cuisine all contribute to a rich tapestry of culinary heritage that reflects the continent's history and cultural depth.

Integrating African Culinary Influences in Contemporary Cooking

For chefs interested in incorporating African culinary influences into their menus, Africa's diverse flavours, ingredients, and techniques provide a wealth of opportunities for creative exploration. Integrating African culinary traditions into contemporary cooking requires not only an understanding of the ingredients and techniques but also a commitment to respecting and honouring the cultural significance of these foods.

One approach to incorporating African influences is to experiment with African spices and seasonings to add depth and complexity to dishes. Spices like berbere (a spicy Ethiopian blend), ras el hanout (a Moroccan blend of over a dozen spices), and alligator pepper (a peppery, aromatic spice from West Africa) offer distinct flavour profiles that can be used to enhance sauces, marinades, and dressings. For example, a chef might add a sprinkle of berbere to roasted vegetables or use ras el hanout to season a slow-cooked lamb dish. These spices add a unique character and warmth to dishes, offering diners a taste of Africa's culinary heritage.

Using traditional African grains and staples can also bring an authentic touch to contemporary menus. Ingredients like millet, sorghum, fonio, and teff have been staples in African diets for centuries, valued for their resilience and nutritional benefits. Chefs can incorporate these grains into salads, pilafs, or even desserts, introducing diners to ancient grains that are both versatile and nutrient-dense. For instance, teff flour can be used to make pancakes or crepes, while fonio, a gluten-free grain from West Africa, can be prepared like couscous and served with vegetables and spices. These ingredients not only add unique textures and flavours but also offer a healthy alternative to more commonly used grains.

Integrating African culinary techniques into contemporary cooking is another way to pay homage to African cuisine. Techniques like slow-cooking stews, grilling over open flames, and fermenting grains allow chefs to recreate the textures and flavours characteristic of African dishes. For example, a chef might prepare a seafood stew inspired by Moroccan tagine, using local fish and vegetables but cooking them slowly in a clay pot to replicate the depth of flavour found in traditional tagine dishes. Alternatively, a chef could explore fermentation by making ogi, a fermented millet or sorghum porridge from Nigeria, and incorporating it into breakfast dishes. Adopting these methods enriches the cooking process, allowing chefs to create dishes that are both flavourful and culturally resonant.

Collaboration with African chefs and culinary experts is an invaluable way to ensure authenticity and deepen one's understanding of African cuisine. Working alongside chefs who specialize in African food offers insights into the cultural significance of certain dishes, helping chefs avoid misrepresentation or superficial adaptations. Collaborative events, guest chef appearances, or cultural exchange programs can bring African culinary traditions to a wider audience, allowing chefs to showcase authentic flavours and techniques while building connections with African food communities. These partnerships foster a spirit of respect and appreciation for African culinary heritage, enriching the dining experience for guests and promoting a deeper cultural understanding.

Incorporating African influences into contemporary cooking also provides chefs with an opportunity to educate diners about African culinary traditions. By sharing the stories behind ingredients and techniques, chefs can create a dining experience that goes beyond taste, offering insight into Africa's diverse food cultures. Menus can include descriptions of traditional dishes, explaining the significance of ingredients like cassava, yam, or baobab, or detailing the origins of dishes like jollof rice or injera. This educational approach not only enhances the dining experience but also helps combat stereotypes and broaden diners' perspectives on African cuisine.

Finally, chefs should approach African culinary traditions with cultural sensitivity and humility, recognizing that African cuisine, like any culinary tradition, is deeply tied to history, identity, and community. Respect for these traditions means honouring their origins and acknowledging the contributions of African cultures to global cuisine. Chefs can do this by sourcing ingredients ethically, studying traditional recipes and techniques, and acknowledging the communities from which these foods originate.

African culinary traditions offer a wealth of flavours, ingredients, and techniques that can inspire contemporary chefs. By exploring African spices, grains, and cooking methods, collaborating with African culinary experts, educating diners, and respecting the cultural significance of these foods, chefs can create dishes that celebrate Africa's rich culinary heritage. Through thoughtful and respectful integration, African influences can bring diversity, depth, and cultural appreciation to modern cuisine.

CHAPTER 11: MIDDLE EASTERN CULINARY HERITAGE

The Rich History Of Middle Eastern Cuisine From Ancient Times To Present

Middle Eastern cuisine is one of the world's oldest and most influential culinary traditions, with roots dating back thousands of years to some of the earliest human civilizations. The region, which includes countries like Egypt, Iran, Iraq, Lebanon, Syria, Turkey, and Israel, has been a crossroads for trade, migration, and cultural exchange. This geographic and cultural diversity has contributed to a rich culinary tapestry, marked by a wide array of flavours, ingredients, and techniques that have shaped the evolution of Middle Eastern food from ancient times to the present.

The Middle East's fertile lands and strategic location along the ancient trade routes, such as the Silk Road and the spice routes, allowed it to become a hub for the exchange of spices, grains, fruits, and culinary knowledge. Early civilizations in Mesopotamia, Egypt, and Persia developed sophisticated agricultural systems that supported a rich diet based on grains, legumes, vegetables, fruits, and animal products. Wheat and barley were staple crops, while chickpeas, lentils, and fava beans provided essential protein. Olive oil, dates, figs, and pomegranates were also abundant and became important ingredients that added complexity to Middle Eastern cuisine.

Middle Eastern cuisine has long placed a strong emphasis on communal dining and hospitality. In ancient Mesopotamia, feasts were often organized to honour gods, celebrate victories, and mark important social occasions. Similarly, the practice of offering food to guests as a gesture of hospitality has been deeply ingrained in Middle Eastern cultures. This tradition continues today, with family-style meals that include a variety of dishes meant for sharing, reflecting the region's emphasis on community and generosity.

The Middle Ages saw the flourishing of the Islamic Golden Age, a period of intellectual, artistic, and culinary innovation that spread from the Middle East to North Africa and Spain. During this time, cookbooks such as the Kitab al-Tabikh were written, documenting recipes and culinary techniques that emphasized flavour balance, presentation, and the use of spices. This era saw the refinement of dishes such as rice pilaf, kebabs, and pastries, many of which were influenced by Persian and Indian cuisines. The spread of Islam also influenced dietary laws, including halal practices, which shaped the ingredients and methods used in

Middle Eastern cooking. The region's culinary tradition is further enriched by its religious and cultural diversity, including Jewish, Christian, and Zoroastrian communities, each of which has contributed its unique practices and dishes.

Modern Middle Eastern cuisine is a blend of ancient traditions and contemporary innovations, with dishes like hummus, falafel, and shawarma enjoyed globally. As Middle Eastern immigrants settled in other countries, they introduced their food to new regions, where these dishes have since been adapted and celebrated. The versatility and adaptability of Middle Eastern cuisine allow it to remain both a cherished tradition and a source of inspiration for modern chefs seeking to explore its complex flavours and techniques.

Key Spices, Techniques, and Dishes (e.g., Kebabs, Flatbreads, Stews)

The flavours of Middle Eastern cuisine are deeply rooted in the use of aromatic spices, herbs, and techniques that bring out the best in each ingredient. A few essential spices and methods, along with iconic dishes like kebabs, flatbreads, and stews, form the heart of Middle Eastern culinary heritage.

Key Spices and Herbs

Middle Eastern cuisine is characterized by its use of spices that add warmth, depth, and subtle complexity to dishes. Sumac, a tangy, slightly sour spice made from dried berries, is used as a seasoning for meats, vegetables, and salads, adding a hint of acidity that balances flavours. Cumin and coriander are staples, often used together to season dishes like kebabs and stews. Za'atar, a blend of wild thyme, sumac, and sesame seeds, is a widely used spice mix sprinkled on breads, vegetables, and meats, lending an earthy and aromatic quality to dishes. Saffron, one of the world's most expensive spices, is used in rice dishes and desserts, adding a unique floral aroma and a deep golden colour. Cardamom, cinnamon, and cloves are often used in sweets and savoury dishes alike, providing a warm, aromatic flavour profile that is quintessentially Middle Eastern.

Traditional Techniques

One of the signature techniques in Middle Eastern cooking is grilling and roasting, especially for meats. Kebabs, for example, are skewered pieces of meat, often marinated with spices, and grilled to perfection. This method imparts a smoky flavour and a charred texture that enhance the spices used in the marinade. Stewing and braising are also central techniques, particularly for dishes like lamb and vegetable stews. Slow-cooking allows the flavours to meld, creating rich, comforting dishes that are deeply flavourful and aromatic. Another important method is baking and stone-oven cooking. Many Middle Eastern households and bakeries use stone ovens to bake flatbreads like pita and lavash, which puff up with an airy texture and a crispy crust when exposed to the high heat.

Fermentation is another key technique, seen in foods like labneh (strained yoghurt) and pickled vegetables. Labneh, a tangy and creamy yoghurt cheese, is often served with olive oil and herbs and is enjoyed as a dip or spread. The practice of pickling vegetables, such as turnips, carrots, and cucumbers, adds acidity and crunch to Middle Eastern meals, providing a refreshing contrast to richer dishes.

Iconic Dishes

Kebabs are perhaps one of the most recognized Middle Eastern dishes, with variations found across the region. Each country has its twist, from shish kebabs in Turkey to kofta kebabs in Lebanon, which are often made with minced meat and spices. Kebabs are traditionally cooked over an open flame, creating a tender and smoky flavour that highlights the spices used.

Flatbreads are another cornerstone of Middle Eastern cuisine. Pita, a soft, round flatbread with a pocket, is commonly used for sandwiches and wraps, filled with falafel, shawarma, or grilled vegetables. Manakish, a flatbread topped with za'atar or cheese, is popular in Levantine countries like Lebanon and Syria. Lavash and barbari are other types of flatbreads found in Persian and Kurdish regions, typically served alongside stews and kebabs.

Stews are a staple in Middle Eastern cuisine, with each country having its variations. Tagines from North Africa are slow-cooked stews made with lamb, chicken, or fish, combined with vegetables, dried fruits, and spices. In Persia (modern-day Iran), ghormeh sabzi— a herbaceous stew made with lamb, kidney beans, and dried limes—is a beloved national dish. Harira, a Moroccan soup with tomatoes, chickpeas, lentils, and spices, is traditionally eaten during Ramadan and exemplifies the comforting flavours of Middle Eastern stews.

These spices, techniques, and dishes are at the heart of Middle Eastern cuisine, reflecting the region's agricultural diversity, rich history, and cultural exchanges. For chefs, these elements offer a wealth of inspiration to create dishes that capture the essence of Middle Eastern cooking.

Applications for Chefs Working with Middle Eastern Flavours and Ingredients

For chefs seeking to incorporate Middle Eastern influences into their menus, the region's bold flavours, spices, and cooking techniques provide numerous opportunities to create dishes that are both contemporary and rooted in tradition. Integrating Middle Eastern flavours into modern cuisine requires an understanding of the ingredients and an appreciation for the cultural significance of these foods.

One approach is to use Middle Eastern spice blends and seasonings to add complexity and warmth to dishes. Spice blends like za'atar, baharat (a Middle Eastern spice mix that often includes black pepper, cumin, coriander, and cinnamon), and ras el hanout (a North African blend with over a dozen spices) can be used to season meats, vegetables, and grains. For instance, za'atar can be sprinkled over roasted vegetables or used as a crust for meats, while baharat can add depth to soups, stews, or even sauces for pasta. These spices introduce a rich, aromatic quality that enhances the flavour of dishes, making them memorable and distinctive.

Incorporating Middle Eastern flatbreads into menus provides a unique way to present appetizers or main courses. Chefs can experiment with freshly baked pita, lavash, or manakish, serving them with a variety of dips like hummus, baba ghanoush, or labneh. Flatbreads can also be topped with innovative ingredients that blend Middle Eastern flavours with local produce, such as a manakish topped with za'atar, goat cheese, and fresh seasonal herbs. These breads serve as a versatile canvas for creativity, allowing chefs to offer guests a taste of Middle Eastern cuisine in a familiar format.

Another way to explore Middle Eastern cuisine is by integrating classic Middle Eastern ingredients into contemporary dishes. Ingredients like pomegranate molasses, sumac, and saffron can add unexpected flavours to familiar dishes. Pomegranate molasses, with its tangy and sweet profile, works beautifully in glazes for meats or as a dressing for salads. Sumac, with its citrusy notes, can replace lemon in marinades, adding a subtle acidity to grilled fish or roasted vegetables. Saffron can be infused into risottos or even used in desserts to add a luxurious, aromatic element that elevates the dish. These ingredients allow chefs to blend Middle Eastern flavours with other cuisines, creating fusion dishes that are innovative yet respectful of tradition.

Showcasing Middle Eastern-style dips and spreads is another way to bring Middle Eastern cuisine to diners. Hummus, baba ghanoush, and muhammara (a dip made from roasted red peppers, walnuts, and pomegranate molasses) are versatile dishes that can be served with flatbreads, crackers, or crudités. Chefs can experiment with variations of these dips, such as adding roasted beetroot to hummus or using charred eggplant in baba ghanoush, to create colourful and flavour-rich options that appeal to a broad audience. These dips are approachable, easily customizable, and perfect for sharing, making them a natural fit for restaurants that emphasize communal dining.

Experimenting with Middle Eastern cooking techniques can also add authenticity and depth to contemporary dishes. Slow-cooking methods, such as those used for tagines, allow chefs to create rich, layered flavours in dishes like braised lamb or chicken. Clay pot cooking, which retains moisture and enhances the flavours of stews and rice dishes, can be used to recreate the texture and depth found in traditional Middle Eastern cuisine. Grilling and roasting, especially with marinades that include ingredients like yoghurt, garlic, and spices, can transform meats, fish, and vegetables, infusing them with Middle Eastern-inspired flavours.

Finally, chefs can enhance the dining experience by educating diners about the history and cultural significance of Middle Eastern cuisine. By including brief descriptions on menus or engaging in conversations with guests, chefs can share the stories behind dishes, ingredients, and techniques. For instance, explaining the origins of za'atar as a traditional spice blend or the cultural significance of sharing a mezze spread helps diners connect with the dishes on a deeper level. This educational approach not only enriches the dining experience but also fosters appreciation for the cultural heritage of Middle Eastern cuisine.

Middle Eastern culinary heritage offers an array of flavours, ingredients, and techniques that can inspire chefs to create dishes that are both innovative and respectful of tradition. By exploring Middle Eastern spices, flatbreads, classic ingredients, and traditional cooking methods, chefs can bring the warmth and richness of Middle Eastern cuisine to modern menus, creating dishes that celebrate this vibrant and ancient food culture.

CHAPTER 12:
THE EUROPEAN
INFLUENCE
ON GLOBAL
GASTRONOMY

European Colonialism And Its Culinary Impact Across Continents

European colonialism, which spanned from the 15th to the 20th centuries, reshaped global cuisine as explorers, merchants, and colonizers introduced European ingredients, cooking techniques, and food preferences to every corner of the world. The exploration and conquest of lands in the Americas, Africa, and Asia not only facilitated the exchange of goods but also dramatically altered food cultures, blending Indigenous ingredients with European culinary practices. This period of colonialism left a complex legacy, as European influences brought about both the enrichment of global gastronomy and the disruption of traditional food systems.

The Age of Exploration, beginning in the 15th century, marked the onset of European expeditions in search of resources, trade routes, and new territories. The Portuguese and Spanish were among the first to establish colonies, soon followed by the Dutch, French, and British. Each European power brought with it a distinct set of foods, cooking methods, and dietary customs that would integrate into, and often overshadow, local food traditions. Colonization enabled the rapid transfer of crops like wheat, barley, sugar, and coffee to the Americas, Africa, and Asia. European settlers established plantations and introduced livestock, such as cattle, pigs, and sheep, to regions where these animals were previously unknown, forever changing the landscapes and diets of local populations.

One of the most notable examples of European influence is the introduction of wheat to the Americas. Native populations in the Americas primarily cultivated corn, beans, and squash, but with the arrival of European colonizers, wheat became a staple. This shift led to the rise of European-style bread across the Americas, particularly in South America, where bread became an essential part of the daily diet. In regions of Africa, European colonists also introduced wheat, which eventually integrated into local diets, resulting in popular dishes like flatbreads and pastries. While bread-making had deep cultural roots in Europe, its adoption across continents reflects the powerful influence European colonists had on local food systems.

European colonial powers also introduced sugarcane

plantations to tropical regions, including the Caribbean, South America, and parts of Asia. The cultivation of sugarcane, driven by European demand for sugar, resulted in the establishment of vast plantations where enslaved Africans were forced to labour, a tragic aspect of culinary history that highlights the ethical complexities of European influence. This period fuelled the global sugar industry, and sugar became central to European culinary traditions, leading to the creation of iconic sweets and pastries like cakes, tarts, and biscuits. The colonial sugar trade had far-reaching effects on global cuisine, as sugar became a valuable and ubiquitous ingredient that transformed both sweet and savoury dishes.

Dairy was another major contribution of European colonists to global food practices. Dairy products, particularly milk, cheese, and butter, were staples in Europe, and as Europeans settled in new regions, they brought livestock and dairy production techniques with them. Cheese-making, virtually unknown in many parts of the world prior to European contact, was introduced to Africa, the Americas, and Asia. Today, dairy products are widely consumed around the world, and cheese, in particular, has become a popular ingredient in global cuisine. European cheese-making techniques inspired local adaptations, such as paneer in India, which was developed after contact with European settlers.

The introduction of European foodways often came at the expense of Indigenous cuisines, as European powers imposed their agricultural practices and eating habits on colonized populations. This colonial legacy continues to shape modern culinary landscapes, as European ingredients and techniques have become integral to global diets. For chefs, understanding the history of European colonial influence on food offers a deeper perspective on the diverse ingredients and techniques that form today's global gastronomy.

Key Ingredients and Techniques Adopted Globally (e.g., Bread-Making, Cheese)

European colonization facilitated the widespread adoption of key ingredients and techniques that are now staples of global cuisine. Among the most influential contributions are bread-making, cheese production, and the cultivation of crops like sugar, potatoes, and coffee.

Bread-Making is one of Europe's most enduring culinary contributions, with roots stretching back to ancient civilizations. Wheat-based breads, in particular, became central to European diets, and as European settlers established colonies, they brought wheat cultivation and bread-making techniques to new lands. Bread took on various forms in different regions, influenced by both European methods and local ingredients. In Latin America, wheat was combined with Indigenous flavours, resulting in unique bread varieties like bolillos in Mexico and pão de queijo in Brazil. In the Middle East, European influence blended with local bread traditions, leading to variations like pita and flatbreads. The French baguette, Italian ciabatta, and British scones are now beloved worldwide, testifying to the global reach of European bread-making traditions.

Cheese production is another hallmark of European cuisine that has left a lasting impact on global culinary practices. The art of cheese-making was perfected in regions like France, Italy, and Switzerland, each of which developed its own styles and techniques based on climate, geography, and local animal breeds. Cheese production spread globally through European trade and colonization, and new varieties emerged as cheese-making techniques were adapted to local tastes and ingredients. For example, Indian cheese-makers adapted the process to create paneer, a fresh, non-melting cheese that became central to many Indian dishes. In Latin America, queso fresco, a mild, crumbly cheese, was developed as a local adaptation of European cheese styles. Cheese has become a universal ingredient, adding flavour, texture, and depth to dishes across cultures.

The introduction of sugar revolutionized both European and global cuisine. Originating in Southeast Asia, sugarcane was brought to Europe by Arab traders and became a prized commodity. The establishment of sugar plantations in the Caribbean and South America by European colonizers transformed sugar into a staple ingredient, driving demand for sweets and leading to the development of iconic European desserts. French pastries, British puddings, and Italian gelato all incorporate sugar as a primary ingredient, and sugar-based desserts became a central part of European culinary traditions. Globally, sugar transformed not only desserts but also savoury dishes, as chefs and home cooks alike began using sugar to balance flavours and enhance sauces.

Potatoes and coffee are two other ingredients that transformed global gastronomy through European influence. Potatoes, native to the Andes in South America, were brought to Europe in the 16th century by Spanish explorers. Though initially met with suspicion, potatoes quickly became a staple crop in Europe, especially in Ireland, Russia, and Germany, where they were essential for feeding growing populations. Potatoes spread to Asia and Africa, where they were incorporated into local cuisines. Today, potatoes are one of the most widely consumed foods in the world, enjoyed in dishes ranging from French fries to Indian aloo gobi.

Coffee, originally from Ethiopia, was introduced to Europe through trade with the Middle East, where coffeehouses had become social and intellectual hubs. The drink quickly gained popularity in Europe, and European colonizers established coffee plantations in tropical regions, particularly in the Caribbean, South America, and Southeast Asia. Coffee became a symbol of European cafe culture, influencing social customs and culinary practices. Today, coffee is one of the world's most consumed beverages, with a diversity of brewing methods and traditions influenced by European coffee culture.

These ingredients and techniques—bread-making, cheese production, sugar, potatoes, and coffee— demonstrate the far-reaching impact of European culinary practices. They have become integral to global cuisine, adapted and reimagined by local cultures, yet retaining their European roots.

How Chefs Can Incorporate European Influences with Cultural Sensitivity

For chefs interested in incorporating European influences into their menus, understanding the historical context of these ingredients and techniques is essential. By acknowledging the legacy of European colonialism and its impact on global food systems, chefs can approach these influences with respect and cultural sensitivity, creating dishes that celebrate diversity while honouring the origins of the ingredients and methods they use.

One approach is to highlight the shared culinary heritage of European and non-European cuisines, emphasizing how ingredients and techniques have evolved through cultural exchange. For example, a chef might create a dish that combines European bread-making techniques with local flavours, such as a baguette topped with Middle Eastern za'atar or an Italian-style flatbread infused with Indian spices. This approach celebrates the global journey of bread as a staple food, acknowledging both its European origins and its adaptation across cultures. Chefs can use their menus to tell the story of these cross-cultural exchanges, helping diners understand how European techniques have been enriched by global influences.

Using European ingredients thoughtfully and respectfully can also enhance the dining experience, especially when working with ingredients that carry historical significance. Sugar, for instance, has a complex colonial history tied to exploitation and forced labour. Chefs might choose to source fair-trade or sustainably produced sugar as a way of honouring the resilience of communities affected by the sugar trade. When creating desserts or baked goods that use sugar as a primary ingredient, chefs can share the history behind these sweets, helping diners appreciate both the pleasure and the cost of this ubiquitous ingredient.

Blending European and local ingredients is another way to incorporate European influences in a way that respects cultural diversity. By combining European techniques with locally sourced ingredients, chefs can create dishes that are uniquely reflective of their environment. For example, a chef might prepare a French-style cheese souffle using a locally made cheese, or create a pasta dish with indigenous herbs and spices. This approach allows chefs to adapt European techniques to local contexts, honouring both the culinary legacy of Europe and the flavours of the local landscape.

Engaging with diverse culinary traditions can also help chefs bring European influences to life in meaningful ways. Collaborating with chefs and food experts from other culinary backgrounds allows for a deeper understanding of how European techniques have been adapted and embraced globally. By inviting guest chefs, hosting cultural exchange events, or attending workshops, chefs can expand their knowledge and appreciation of global cuisine, ensuring that their interpretations of European dishes remain authentic and respectful.

Education and transparency play a key role in incorporating European influences with cultural sensitivity. Chefs can educate diners about the origins and historical significance of the dishes they prepare, providing context that enriches the dining experience. For example, a chef might introduce a dish inspired by British pub fare, such as fish and chips, but offer a brief description on the menu that explains its cultural background. Similarly, when preparing dishes with ingredients like potatoes or tomatoes, chefs can mention their journey from the Americas to Europe, highlighting the global movement of ingredients that shaped modern cuisine. This educational approach fosters cultural appreciation and deepens diners' connection to the food they are enjoying.

Finally, approaching European culinary traditions with humility and respect is essential. European cuisine has a rich history of regional diversity, from the butter-rich dishes of northern France to the olive oil-laden fare of the Mediterranean. Chefs should avoid reducing European cuisine to stereotypes, instead exploring the unique flavours and techniques that define each European region. This nuanced approach allows for a more comprehensive understanding of European culinary heritage, emphasizing the diversity within Europe itself and encouraging chefs to explore its flavours thoughtfully.

European influences on global gastronomy are both profound and complex, offering chefs a wealth of ingredients, techniques, and historical context to draw from. By understanding the legacy of European colonialism, celebrating the journey of key ingredients and techniques, and approaching European influences with cultural sensitivity, chefs can create menus that honour this rich heritage. This respectful and thoughtful approach to incorporating European elements allows for a dining experience that is both flavourful and meaningful, highlighting the interconnectedness of global cuisine and the shared history that has shaped it.

CHAPTER 13: CULINARY TRADITIONS OF SOUTH ASIA

Overview Of India, Pakistan, Sri Lanka, And Bangladesh's Culinary Traditions

The cuisine of South Asia, encompassing India, Pakistan, Sri Lanka, and Bangladesh, is among the world's most diverse and flavourful. Each country has its unique culinary traditions shaped by geography, climate, religious practices, and historical influences. These traditions offer a broad spectrum of flavours, textures, and techniques that have evolved over centuries, reflecting a harmonious blend of spices, ingredients, and cooking methods that embody the cultural richness of the region.

India's culinary landscape is vast and varied, characterized by regional diversity that results in distinct culinary styles across its states. Northern India, influenced by Mughal cuisine, is known for its rich, creamy gravies, often featuring spices like cardamom, cinnamon, and cloves, as well as the use of dairy in the form of ghee, yoghurt, and paneer. Dishes like biryani, rogan josh, and butter chicken exemplify this style. Southern Indian cuisine, by contrast, is known for its spicier flavours, extensive use of coconut, tamarind, and curry leaves, and a focus on rice and lentils. Dishes like dosa, sambar, and rasam are staples in the south, reflecting a reliance on local ingredients and vegetarian influences. Eastern India, particularly Bengal, is renowned for its seafood and sweets, such as macher jhol (fish curry) and rasgulla (a popular milk-based dessert). Western India, including Gujarat and Maharashtra, features vegetarian-friendly dishes and complex flavours with a blend of sweetness and spice, as seen in dishes like dhokla and vada pav.

Pakistan's cuisine shares similarities with Northern Indian cuisine, especially in the use of spices, but has its own identity marked by a focus on meats, especially lamb, beef, and goat. The cuisine reflects the country's Islamic heritage, with dishes like biryani, kebabs, and haleem—a rich, slow-cooked dish made with meat, lentils, and wheat. Punjabi cuisine, known for its hearty flavours and use of tandoori methods, is popular across Pakistan, while Karachi's street food culture and Balochistan's simpler, meat-centric dishes showcase the diversity within the country. The use of spices like garam masala, cumin, and coriander, as well as cooking techniques such as grilling and slow-cooking, define the bold and robust flavours of Pakistani cuisine.

Sri Lankan cuisine is deeply influenced by its tropical environment and proximity to the Indian Ocean, resulting in a diet rich in seafood, coconut, and rice. Known for its liberal use of spices, Sri Lankan food often combines cinnamon, cloves, cardamom, and curry leaves, creating a complex, aromatic flavour profile. Coconut milk is frequently used in curries, and dried fish is a common ingredient, adding a savoury depth to dishes. Popular dishes include rice and curry, hoppers (a type of fermented rice pancake), and sambols (spicy relishes often made with coconut or chili). Sri Lankan cuisine reflects influences from South Indian, Dutch, and Malay cooking, highlighting the island's role as a cultural crossroads.

Bangladeshi cuisine is heavily influenced by Bengal's rivers and fertile plains, resulting in a diet centred around rice and fish. Known as the "land of rivers," Bangladesh boasts an impressive variety of freshwater fish, such as hilsa, which is often marinated with mustard paste and steamed in banana leaves. Bengali cuisine, shared across parts of Bangladesh and Eastern India, is famous for its delicate use of spices, often with a slight sweetness, and the use of mustard oil, which adds a pungent flavour to dishes. Lentils, vegetables, and rice are staples, with iconic dishes like bhuna (a slow-cooked curry), bhorta (mashed vegetables or fish with spices), and panta bhat (fermented rice served with accompaniments). Desserts like mishti doi (sweet yoghurt) and roshogolla are also popular, showcasing the region's penchant for sweets.

Together, the culinary traditions of India, Pakistan, Sri Lanka, and Bangladesh represent a tapestry of flavours, techniques, and ingredients that embody the cultural and geographic diversity of South Asia. Each country's cuisine is rooted in local ingredients and influenced by historical trade, migration, and colonialism, making South Asian food a fascinating study of adaptation and innovation.

Case Studies of Iconic Dishes and Flavour Profiles

The distinctive flavour profiles of South Asian cuisine are showcased in a range of iconic dishes, each of which exemplifies the region's culinary creativity and mastery of spices. By examining a few of these dishes, chefs can gain insight into the techniques and ingredient combinations that define South Asian cooking.

Biryani is one of the most celebrated dishes across South Asia, with variations found in India, Pakistan, and Bangladesh. Originally brought to the region by Persian traders and Mughal emperors, biryani is a fragrant rice dish that combines basmati rice, meat (such as chicken, lamb, or goat), and a medley of spices, including cardamom, cloves, and saffron. In India, Hyderabad's biryani is famous for its aromatic spices and use of fried onions, while Pakistan's Sindhi biryani is known for its spiciness and inclusion of tomatoes. Bangladeshi biryani, often served with spicy pickles, showcases a unique regional twist. The layering of flavours and textures, along with the use of slow-cooking to allow the spices to permeate the rice and meat, makes biryani an exemplary dish of South Asian cuisine's complexity.

Butter Chicken (Murgh Makhani), a beloved dish from Northern India, is a rich, creamy curry made with marinated chicken cooked in a tomato-based sauce with butter and cream. The dish's origin is attributed to the chefs of Moti Mahal restaurant in Delhi, who developed it as a way to use leftover tandoori chicken by simmering it in a spiced gravy. Butter chicken's mild flavour, balanced by the tang of tomatoes and the richness of butter and cream, has made it popular worldwide. The use of tandoori techniques to prepare the chicken, followed by slow-cooking in a sauce, is a hallmark of Northern Indian cooking, which emphasizes robust flavours and creamy textures.

Hoppers (Appam) from Sri Lanka are a unique dish that highlights the island's use of fermented ingredients. Made from a batter of rice flour and coconut milk, hoppers are thin, bowl-shaped pancakes with a crispy edge and a soft, spongy centre. They are often served with sambols, chutneys, or curry, creating a perfect balance of texture and flavour. The fermentation process, which gives hoppers their distinctive sour flavour, reflects the influence of tropical ingredients and techniques that add depth to Sri Lankan cuisine. This dish exemplifies Sri Lankan ingenuity in using local ingredients to create a versatile, satisfying food that can be enjoyed at any meal.

Bengali Fish Curry (Macher Jhol) from Bangladesh and Eastern India is a dish that showcases the region's use of mustard oil and fresh fish, two ingredients central to Bengali cooking. Made with hilsa or other freshwater fish, macher jhol is a light yet flavourful curry with mustard paste, turmeric, green chilies, and fresh coriander. The mustard oil lends a sharp, pungent flavour, which, combined with turmeric and spices, creates a bold yet balanced taste. This dish exemplifies the Bengali philosophy of letting the main ingredient —often fish—shine through with minimal spices, a contrast to the heavier curries of Northern India.

Each of these dishes showcases the complexity and diversity of South Asian cuisine, where spices, techniques, and local ingredients combine to create distinctive and memorable flavours. For chefs, studying these iconic dishes offers insights into the meticulous layering of spices, the importance of balance, and the role of local ingredients in South Asian cooking.

How Chefs Can Adapt South Asian Techniques to Various Cuisines

South Asian cooking offers a wealth of techniques and ingredients that chefs can adapt to enhance the flavours and textures of their own dishes. By incorporating South Asian methods with a creative approach, chefs can create dishes that pay homage to the region's culinary heritage while appealing to a global audience.

One of the most valuable techniques for chefs to adopt is the art of spice blending. South Asian cuisine relies on complex spice blends, such as garam masala in India and Pakistan or the Sri Lankan spice mix that often includes cinnamon, cardamom, and fenugreek. Chefs can create their custom spice blends by experimenting with the proportions and combinations of spices, using them to add depth to soups, stews, and marinades. For example, a chef might add a pinch of garam masala to a traditional Western stew, introducing a warm, aromatic quality that complements the dish's existing flavours. By mastering spice blending, chefs can enhance the flavour profile of their dishes, bringing an element of South Asian sophistication to diverse culinary traditions.

Marination and tenderization techniques used in South Asian cuisine are also valuable for adapting to different proteins and flavours. Tandoori marinades, which often combine yoghurt, garlic, ginger, and spices, can be used to marinate not only chicken but also seafood, lamb, or even vegetables. The yoghurt acts as a tenderizer, breaking down the protein while infusing it with spices, resulting in a flavourful, juicy final product. Chefs might experiment with this technique by marinating salmon or shrimp in a tandoori-inspired mixture, then grilling or baking it for a unique fusion dish that balances tenderness with robust seasoning.

Incorporating South Asian sauces and chutneys into Western dishes is another way for chefs to blend culinary traditions. Mint chutney, tamarind sauce, and coconut chutney are versatile condiments that add vibrant flavours to a variety of dishes. Chefs can use these sauces as accompaniments for grilled meats, as dressings for salads, or even as dipping sauces for appetizers. Tamarind sauce, for example, with its sweet and tangy profile, can complement dishes ranging from roasted meats to vegetable sides. Mint chutney, with its refreshing and herbaceous flavour, can be used as a dressing for grain bowls or drizzled over roasted potatoes, bringing a South Asian twist to familiar dishes.

For chefs interested in exploring vegetarian cuisine, South Asian vegetarian techniques offer an abundance of inspiration. Dishes like dal (lentil stew), aloo gobi (potato and cauliflower curry), and saag paneer (spinach and cheese) are flavourful and filling, showcasing the variety that can be achieved with plant-based ingredients. Chefs can experiment with creating hearty, spice-rich vegetarian dishes by incorporating these techniques and ingredients. For example, lentils can be slow-cooked with garlic, ginger, and a blend of spices to create a comforting, protein-rich stew, while vegetables can be sautéed with mustard seeds and curry leaves for a burst of South Asian flavour. These techniques provide chefs with tools to create satisfying vegetarian options that are as rich and complex as meat-based dishes.

Fusion dishes that combine South Asian elements with other culinary traditions allow chefs to innovate while respecting cultural origins. For instance, a chef might create a pizza with a tikka masala sauce base, topped with paneer, red onions, and fresh cilantro, merging Italian and Indian flavours. Similarly, a sushi roll could incorporate South Asian flavours by adding a touch of mango chutney, mint, or chili to the filling, creating a fusion that balances the fresh taste of sushi with the vibrant spices of South Asia. These fusion dishes offer diners a novel experience that celebrates the adaptability and versatility of South Asian ingredients and techniques.

Finally, chefs can enhance their approach to plating by drawing inspiration from South Asia's colourful and vibrant presentation. South Asian cuisine often features a variety of colours, textures, and garnishes, which add visual appeal to the dining experience. Chefs can incorporate colourful ingredients, such as turmeric-infused rice, purple eggplants, or bright red chilies, into their presentations, creating dishes that are visually as well as flavourfully engaging. Garnishes like fresh coriander, thinly sliced red onions, or a sprinkle of pomegranate seeds can add texture and brightness to plates, transforming the dining experience into a feast for the senses.

South Asian cuisine offers a wealth of techniques, flavours, and ingredients that chefs can adapt across a variety of culinary contexts. By experimenting with spice blends, marination methods, sauces, vegetarian dishes, fusion creations, and vibrant presentation, chefs can bring the warmth and richness of South Asia to their kitchens, creating dishes that honour both tradition and innovation. This respectful and creative approach allows chefs to celebrate South Asia's culinary heritage while adding their unique touch to its timeless flavours.

CHAPTER 14: THE RISE OF FUSION CUISINE

Origins Of Fusion Cuisine And Its Place In Culinary Anthropology

Fusion cuisine, a culinary approach that combines elements from different culinary traditions, has a rich and multifaceted history, with origins that trace back to ancient trade routes and cross-cultural exchanges. While fusion cooking is often seen as a modern trend, it is deeply rooted in the anthropological evolution of food, reflecting centuries of exploration, migration, and adaptation. As people and cultures interacted throughout history, ingredients, cooking techniques, and flavour profiles were exchanged and adapted, resulting in new and unique culinary traditions.

The Silk Road is one of the earliest examples of cross-cultural culinary exchange, where merchants and travellers from Asia, the Middle East, and Europe traded spices, herbs, and cooking techniques. Ingredients like saffron, pepper, and cinnamon spread along these routes, making their way into the kitchens of different civilizations and leading to the blending of flavours and dishes. Similarly, the Age of Exploration in the 15th and 16th centuries furthered culinary fusion as European explorers brought back ingredients like potatoes, tomatoes, and chilies from the Americas, which were then incorporated into European, Asian, and African cuisines. The spread of these new ingredients transformed traditional recipes and laid the foundation for early fusion dishes, with chili peppers becoming central to Asian cuisines and tomatoes transforming Italian and Mediterranean dishes.

The modern concept of fusion cuisine, however, began to take shape in the late 20th century, when chefs started to experiment more deliberately with combining distinct culinary traditions. The globalized world allowed chefs to access diverse ingredients, inspiring them to blend traditional flavours in innovative ways. The 1980s and 1990s saw the rise of Pan-Asian and Californian fusion, as chefs in places like Los Angeles, San Francisco, and New York began combining flavours from Asian, Latin American, and European cuisines. Restaurants became experimental spaces, where chefs took inspiration from multiple traditions to create unique dishes that reflected a world increasingly interconnected by travel, migration, and trade.

In culinary anthropology, fusion cuisine is a powerful reflection of cultural blending and adaptation, where food becomes a medium through which people can experience and understand other cultures. At its best, fusion cuisine highlights commonalities between culinary traditions and introduces diners to new flavours and textures. However, the rise of fusion cuisine has also raised questions about cultural appropriation and authenticity, as chefs must navigate the fine line between creativity and respect for the culinary heritage of different cultures. For chefs, understanding the origins of fusion cuisine in both historical and modern contexts is essential for creating dishes that honour the integrity of each culinary tradition while celebrating innovation.

Famous Fusion Dishes and the Cultural Exchange They Represent

Fusion cuisine has given rise to some of the world's most beloved dishes, each a testament to cultural exchange and culinary creativity. These dishes reflect how different food traditions can come together, creating something unique that bridges distinct flavours, ingredients, and techniques.

Sushi Burrito is a modern fusion dish that combines Japanese and Mexican culinary elements. This creation consists of sushi ingredients, such as raw fish, rice, and seaweed, wrapped in a larger burrito-style roll. The sushi burrito reflects the popularity of both sushi and burritos in Western markets and is an example of how chefs adapted Japanese flavours to appeal to American and Mexican-inspired food culture. The dish maintains the freshness and flavours of sushi but incorporates the convenient, hand-held form of a burrito, creating a fusion that is as practical as it is flavourful. The sushi burrito has become a global phenomenon, often featuring innovative fillings like spicy tuna, avocado, and kimchi, showcasing the versatility of fusion cuisine.

Tandoori Chicken Pizza blends Indian and Italian culinary traditions, combining tandoori-spiced chicken with a classic pizza base, topped with ingredients like cilantro, red onions, and sometimes a drizzle of yoghurt. This dish reflects the growing popularity of Indian flavours in Western cuisine, while using pizza, a universally beloved food, as a canvas for experimenting with spices like cumin, coriander, and turmeric. Tandoori chicken pizza is a favourite in places like the United Kingdom and the United States, where Indian communities have influenced local food culture. This dish represents how fusion cuisine can be both accessible and bold, introducing diners to the depth of Indian flavours within a familiar format.

Korean Tacos are a famous example of fusion that brings together Korean and Mexican flavours. Created by Chef Roy Choi in Los Angeles, Korean tacos typically feature marinated beef (bulgogi) served in a soft tortilla and topped with kimchi, sesame seeds, and a spicy sauce. The Korean taco reflects the multicultural landscape of Los Angeles, where Mexican and Korean communities coexist, creating opportunities for culinary blending. The bulgogi marinade, with soy sauce, ginger, and garlic, melds seamlessly with the Mexican tortilla, while the kimchi adds a tangy, spicy contrast. Korean tacos have become an iconic dish in the United States, celebrated for their vibrant flavours and unique cultural symbolism.

Ramen Burger combines Japanese and American

flavours, featuring a hamburger patty sandwiched between two ramen noodle "buns." This dish, created by Chef Keizo Shimamoto in New York, exemplifies the playfulness of fusion cuisine, transforming familiar ingredients in unexpected ways. The ramen burger preserves the flavours of a traditional burger, with toppings like lettuce, cheese, and special sauce, but uses crispy, pan-fried ramen noodles in place of a standard bun. This fusion is particularly popular among younger diners, who appreciate its novelty and the way it reimagines a traditional American dish with Japanese elements. The ramen burger represents the global influence of Japanese cuisine, which has become a source of inspiration for chefs around the world.

Bánh Mì Sandwich is a Vietnamese fusion dish with French influences, resulting from the period of French colonial rule in Vietnam. The sandwich features a French baguette filled with Vietnamese ingredients, such as pickled carrots, cucumbers, cilantro, pâté, and a variety of meats. The bánh mì is a perfect example of how colonial history can shape a country's culinary traditions, combining the French technique of bread-making with Vietnamese flavours. Today, bánh mì is a popular street food in Vietnam and has gained international fame, representing a harmonious blend of two distinct culinary traditions.

These dishes exemplify the power of fusion cuisine to bring together diverse flavours, highlighting the ways in which cultural exchange has shaped global gastronomy. Each fusion dish represents a unique culinary story, inviting diners to experience the intersection of different cultures in a single bite.

Advice for Chefs on Creating Respectful, Innovative Fusion Dishes

Creating successful fusion dishes requires a delicate balance between innovation and respect for the culinary traditions being combined. As fusion cuisine grows in popularity, chefs have a responsibility to approach it with cultural sensitivity, ensuring that the dishes they create honour the integrity of each cuisine while introducing new and exciting flavours.

1. Research and Understand the Origins of Each Cuisine

The first step in creating respectful fusion dishes is to thoroughly research the cuisines being combined. This includes understanding the historical, cultural, and social significance of key ingredients, dishes, and techniques. For example, if a chef wishes to combine Japanese and Mexican flavours, it's important to understand the role of umami in Japanese cooking and the balance of acidity, heat, and spices in Mexican cuisine. By understanding the fundamentals of each culinary tradition, chefs can create dishes that celebrate, rather than dilute, these unique elements. A deep knowledge of each cuisine allows for more thoughtful and authentic fusion creations.

2. Preserve the Integrity of Key Ingredients and Techniques

Fusion cuisine can sometimes lead to superficial adaptations that overlook the essence of a dish. To avoid this, chefs should preserve the integrity of key ingredients and techniques that are central to each cuisine. For instance, if creating a dish that combines Italian and Thai flavours, a chef might incorporate Thai basil and fish sauce into a classic Italian pasta recipe. However, they should be mindful to maintain the traditional Italian technique of preparing pasta, rather than compromising the texture or method. By preserving core techniques, chefs can create fusion dishes that are innovative yet remain true to the culinary principles of each cuisine.

3. Embrace Balance and Harmony in Flavour Profiles

South Asian cuisine, for example, might emphasize spices, while East Asian cuisine might focus on subtle umami. Finding a balance between these flavours ensures that neither cuisine overpowers the other, allowing the fusion dish to highlight the best aspects of each tradition. Fusion dishes should avoid overwhelming flavours or overly complicated combinations, as simplicity often allows the unique qualities of each ingredient to shine. For example, a chef creating a fusion curry with Mediterranean elements might combine Indian spices with a tomato and olive base, achieving harmony without overwhelming the dish.

4. Experiment with Complementary Ingredients and Textures

Successful fusion cuisine often involves pairing complementary ingredients that bring out the best in each other. Chefs should experiment with ingredients from different cuisines that share similar textures or flavour profiles. For instance, coconut milk from Thai cuisine pairs well with Italian basil and tomatoes, as both cuisines use ingredients that create creamy, flavourful dishes. Similarly, a chef might experiment with the textures of Mexican masa (corn dough) and Japanese rice flour to create a fusion dumpling. Exploring complementary ingredients allows chefs to create fusion dishes that feel natural and balanced, enhancing the dining experience.

5. Collaborate with Chefs and Experts from Different Culinary Backgrounds

One of the most effective ways to create respectful fusion dishes is through collaboration with chefs from diverse culinary backgrounds. Working alongside chefs who specialize in different cuisines allows for a deeper understanding of the flavours, techniques, and cultural significance of each culinary tradition. Collaboration can lead to more authentic fusion dishes, as chefs gain insights that they might not have otherwise considered. For example, a chef trained in French cuisine might collaborate with a Korean chef to create a fusion dish that combines the meticulous plating of French dishes with the bold flavours of Korean barbecue.

6. Educate Diners About the Fusion Dish's Origins

Educating diners about the origins and cultural significance of a fusion dish can add depth to their dining experience. By including descriptions on the menu or engaging in conversations with guests, chefs can share the inspiration behind their fusion creations. For example, a chef might explain the history of the bánh mì, noting how French and Vietnamese culinary traditions combined to create this beloved sandwich.

Educating diners not only enhances their appreciation of the dish but also encourages cultural respect and awareness.

7. Approach Fusion with Respect and Sensitivity to Cultural Contexts

Finally, chefs should approach fusion cuisine with a sense of humility and respect. Fusion cuisine has the potential to bridge cultures, but it can also be perceived as insensitive if executed poorly. Chefs should avoid using elements of a cuisine purely for novelty or aesthetic purposes, as this can lead to cultural appropriation. Instead, fusion dishes should be created with a genuine appreciation for the culinary traditions involved, honouring the significance of each ingredient and technique.

Fusion cuisine is a celebration of global culinary diversity, offering chefs the opportunity to create dishes that reflect the interconnectedness of cultures. By approaching fusion with respect, understanding, and a commitment to authenticity, chefs can create innovative dishes that honour the culinary heritage of each tradition. This thoughtful approach not only elevates the art of fusion cuisine but also fosters a deeper appreciation for the world's diverse food cultures.

CHAPTER 15: THE INFLUENCE OF COLONIALISM ON FOOD SYSTEMS

Overview Of Colonial Food Systems And Forced Crop Cultivation

The colonial era, spanning from the 15th to the early 20th centuries, marked a period of profound transformation in global food systems. As European powers established colonies across Asia, Africa, and the Americas, they imposed new agricultural practices, introduced non-native crops, and exploited both land and labour to meet the demands of European markets. These colonial food systems were driven by profit, leading to forced cultivation and the disruption of Indigenous agricultural practices, which had sustained communities for centuries. The legacy of these systems continues to influence contemporary food production and consumption patterns.

One of the central features of colonial food systems

was the establishment of monoculture plantations, where a single crop was cultivated on a massive scale. Cash crops like sugarcane, cotton, coffee, tea, and tobacco were prioritized, as they were highly profitable and in high demand in Europe. Indigenous crops and diverse agricultural systems were often replaced by these monocultures, reducing local food security and rendering colonies dependent on imported foods. For instance, the Caribbean islands, once home to a diversity of Indigenous crops, became dominated by sugarcane plantations, while tea plantations proliferated in India and coffee plantations spread across Central and South America.

These monoculture systems required a labour force to sustain them, which led to the forced migration and enslavement of millions of people. Africans were transported across the Atlantic to work on sugarcane, cotton, and tobacco plantations in the Americas, often under brutal conditions. Similarly, the British Empire transported Indian indentured labourers to work on sugar plantations in the Caribbean and Mauritius. This exploitation of labour for colonial profit contributed to a system that prioritized economic gain over the well-being of people and the environment.

Colonial powers also introduced new crops to their colonies, often with little regard for local diets or ecological balance. These crops, chosen primarily for their economic value rather than their suitability to the environment, frequently disrupted Indigenous ecosystems and practices. The introduction of maize, potatoes, and cassava to Africa, for instance, transformed diets and agricultural practices, as these crops became staples. While these crops provided nutritional benefits, the imposed agricultural changes often marginalized traditional farming practices, causing a loss of agricultural biodiversity and cultural identity.

Colonial food systems, therefore, were designed to serve the economic interests of the colonizers, often at the expense of local communities and ecosystems. The imposition of monoculture, forced labour, and non-native crops had lasting effects on the food sovereignty of colonized regions, many of which remain food insecure today. For chefs and culinary professionals, understanding the legacy of these colonial systems is essential for recognizing the historical injustices embedded within modern food systems and considering ways to support food sovereignty and sustainability in their practices.

Examples of Colonial Impact in Asia, the Americas, and Africa

The impact of colonialism on food systems can be seen through specific examples in Asia, the Americas, and Africa, each of which illustrates the profound changes imposed by European powers.

Asia: The Tea and Opium Trade in India and China

In Asia, British colonial rule reshaped agricultural practices, particularly in India, where the British East India Company established extensive tea plantations. Before British colonization, tea was primarily grown and consumed in China. However, as British demand for tea grew, the British established their own tea plantations in the Darjeeling and Assam regions of India. Indigenous forests were cleared, and local communities were often displaced or coerced into working on these plantations under harsh conditions. This focus on tea cultivation replaced other crops that had traditionally sustained Indian communities, contributing to food insecurity.

In China, British influence in trade culminated in the opium trade, which, although not a food crop, had devastating impacts on Chinese society. The British traded Indian-grown opium for Chinese tea, leading to widespread addiction in China and prompting the Opium Wars. This trade had far-reaching social and economic consequences for China, disrupting local agriculture and social structures. These examples highlight how British colonialism prioritized profit over the well-being of local populations, reshaping agricultural systems and food practices in both India and China.

The Americas: Sugar Plantations in the Caribbean and Central America

The Caribbean islands and parts of Central America became central to the production of sugar, which was in high demand in Europe. The establishment of sugarcane plantations transformed the Caribbean landscape, displacing Indigenous crops and severely impacting local food sovereignty. These plantations relied heavily on enslaved Africans, who were forced to work in inhumane conditions. The cultivation of sugar required intense labour, and enslaved workers often faced brutal treatment, long hours, and poor living conditions.

The legacy of sugar plantations continues to shape the Caribbean's economy and food culture. Today, the Caribbean remains heavily reliant on imported foods, as the colonial emphasis on cash crops limited the region's agricultural diversity. While sugar remains an integral part of Caribbean cuisine and culture, the painful history of its production highlights the social and economic inequalities that underpinned colonial food systems. In Latin America, coffee plantations similarly prioritized export crops over local food security, leading to a reliance on cash crops that persists to this day.

Africa: Cocoa and Palm Oil in West Africa

In West Africa, colonial powers, particularly the British and the French, established plantations for cocoa and palm oil, two crops that were highly valuable in European markets. Cocoa, now a staple of the global chocolate industry, was introduced to West Africa by European colonists who established vast cocoa plantations in Ghana and Côte d'Ivoire. While cocoa production has provided economic opportunities for the region, it has also led to issues such as child labour, deforestation, and food insecurity, as land is devoted to cash crops rather than local food production.

Palm oil, native to West Africa, was exploited on a massive scale during the colonial period and remains a major export today. European demand for palm oil, used in both food products and industrial applications, spurred the establishment of large-scale palm plantations, displacing local communities and reducing biodiversity. The legacy of colonial palm oil production continues to affect West African countries, where small farmers struggle against large agribusinesses, and monoculture palm plantations threaten ecosystems.

These examples illustrate the widespread impact of colonialism on food systems in Asia, the Americas, and Africa. Colonial powers imposed agricultural practices that prioritized profit, often at the expense of local communities, ecosystems, and food security. The legacy of these practices persists in modern food systems, as many former colonies continue to grapple with the economic, environmental, and social consequences of colonial exploitation.

Reflections for Chefs on Food Sovereignty and Ethical Sourcing

For chefs and culinary professionals, understanding the colonial impact on food systems offers an opportunity to engage with issues of food sovereignty and ethical sourcing. Food sovereignty, a concept championed by social movements such as La Via Campesina, emphasizes the right of communities to control their own food systems, including what crops they grow and how food is produced and distributed. By considering food sovereignty and sourcing practices, chefs can play a role in supporting sustainable, fair, and just food systems.

1. Support Indigenous and Local Food Producers

One way chefs can promote food sovereignty is by sourcing ingredients from Indigenous and local food producers. Many communities are working to reclaim their traditional food systems, cultivating crops that reflect their cultural heritage and local ecosystems. By purchasing ingredients from these producers, chefs support the revival of Indigenous agricultural practices and help foster economic independence for marginalized communities. For instance, a chef might source wild rice from Native American farmers in the United States or purchase traditional maize varieties from Indigenous farmers in Mexico, helping to sustain these culturally significant crops.

2. Choose Fair Trade and Ethically Sourced Ingredients

For crops with colonial histories, such as coffee, sugar, cocoa, and tea, chefs can make a positive impact by choosing fair trade and ethically sourced options. Fair trade certification ensures that farmers receive fair wages and work under humane conditions, addressing some of the exploitation embedded in colonial food systems. By choosing fair trade coffee or chocolate, chefs contribute to ethical food practices that respect the labour of small farmers. Additionally, chefs can educate their diners about the importance of ethical sourcing, raising awareness of the historical and social context of these ingredients.

3. Celebrate Cultural Resilience Through Traditional Ingredients

Colonialism often marginalized Indigenous ingredients and cuisines, but many communities have preserved their culinary heritage despite these challenges. Chefs can celebrate this resilience by incorporating traditional ingredients into their menus and honouring their significance. For example, a chef might feature cassava, a traditional crop in many African and South American cuisines, or use teff flour to create dishes that celebrate Ethiopian culinary traditions. This approach allows chefs to highlight the cultural value of these ingredients, honouring the resilience of communities that have preserved their foodways in the face of colonial pressures.

4. Practice Transparency and Educate Diners

Transparency is key to fostering ethical food practices. Chefs can engage diners in conversations about the origins of their ingredients, explaining the historical significance of crops like sugar, coffee, or cocoa. By including information on menus or engaging in discussions with guests, chefs help diners understand the complexities of global food systems and encourage them to make informed choices. This approach not only enriches the dining experience but also promotes a greater awareness of food sovereignty and the impact of colonialism on today's food landscape.

5. Advocate for Sustainable Agriculture and Biodiversity

Colonial monocultures reduced agricultural biodiversity, leaving many regions dependent on a narrow range of crops. Chefs can counteract this legacy by supporting sustainable agriculture practices that prioritize biodiversity. Sourcing ingredients from small farms, promoting heirloom varieties, and emphasizing seasonal produce are all ways chefs can contribute to biodiversity and sustainability. By creating menus that highlight diverse ingredients and rotate with the seasons, chefs encourage a more sustainable approach to food that honours the environment and reduces reliance on monoculture systems.

6. Engage in Continuous Learning and Reflection

Finally, chefs can deepen their engagement with food sovereignty and ethical sourcing by continuously learning about the historical and social context of the ingredients they use. Colonialism's impact on food systems is complex, and understanding its legacy requires a commitment to education and reflection. Chefs might read about the history of global food systems, attend workshops on food justice, or collaborate with organizations that advocate for food sovereignty. This commitment to learning fosters a more thoughtful approach to cooking, encouraging chefs to use their influence to support justice and equity in the food industry.

Colonialism has left an indelible mark on global food systems, shaping the way ingredients are produced, distributed, and consumed. For chefs, understanding this history is essential for creating menus that reflect ethical sourcing and support food sovereignty. By prioritizing fair trade, supporting local producers, celebrating traditional ingredients, and educating diners, chefs can contribute to a food system that respects both cultural heritage and environmental sustainability. This commitment to ethical practices helps honour the resilience of communities impacted by colonialism, creating a culinary experience that is both delicious and meaningful.

CHAPTER 16: URBANIZATION AND STREET FOOD CULTURE

Origins Of Street Food And Its Role In Urban Communities

Street food has a long history, with origins that can be traced back thousands of years to some of the earliest urban civilizations. Its role in cities, particularly in densely populated urban areas, has always been significant, providing accessible, affordable, and flavourful meals to people on the move. Street food is more than just a convenient dining option; it reflects local culture, tradition, and community. As cities grew and urbanization transformed the way people lived and worked, street food adapted to meet the needs of diverse urban populations, evolving into a vibrant and essential component of city life.

The earliest records of street food date back to ancient Greece and Rome, where vendors sold small, ready-to-eat meals to city dwellers who didn't have kitchens at home. In Greece, vendors sold fish and bread, while in Rome, food stalls offered dishes like chickpea pancakes and fried fish. Street food allowed busy workers to eat on the go, ensuring that even those with little time or money could have access to hot, freshly made food. This model continued to thrive across various ancient civilizations, from Egypt to China, where street food vendors would set up near busy marketplaces or along major roads, catering to traders, labourers, and travellers.

As urbanization accelerated in the modern era, street food became an increasingly integral part of city life. In many parts of the world, particularly in Asia, Africa, and Latin America, street food evolved into an essential component of urban culture. Vendors sold traditional foods that were affordable, easy to eat, and rich in local flavour, often drawing on family recipes passed down through generations. Street food quickly became associated with community identity, as each region's unique flavours and techniques shaped the types of food available on its streets. In cities like Bangkok, Mumbai, and Mexico City, street food grew to represent the culinary soul of the community, offering dishes that were authentic, accessible, and reflective of local tastes and customs.

Street food's significance lies not only in its flavours but also in its social and economic impact. Street vendors provide affordable food options for urban populations, especially for working-class residents and low-income communities. They offer quick and affordable meals to students, office workers, and travellers, creating a sense of inclusivity in an often expensive urban dining landscape. Street food vendors also play a key role in local economies, often operating family-run businesses that contribute to economic activity and community livelihoods. Moreover, street food reflects cultural diversity within cities, as migrants bring their culinary traditions with them, enriching the street food scene with flavours from different regions and countries.

In the modern era, street food has gained international recognition as an exciting culinary experience. Food tourism has grown significantly, with travellers seeking out street food markets and stalls as a way to experience the authentic flavours of a place. Today, street food is celebrated for its creativity, cultural authenticity, and ability to bring people together, making it a symbol of urban culture and culinary diversity.

Exploration of Famous Street Food Hubs (e.g., Bangkok, Istanbul, Mexico City)

Several cities around the world are renowned for their vibrant street food culture, each offering a unique experience that reflects the culinary heritage and diversity of the region. Exploring these hubs provides insight into how street food can define a city's identity and offer a window into its cultural heart.

Bangkok, Thailand

Bangkok is one of the world's most celebrated street food capitals, known for its diverse array of flavours, aromas, and textures. Thai street food reflects a balance of sweet, sour, salty, and spicy flavours, creating dishes that are both satisfying and complex. Bangkok's street food scene is vibrant and dynamic, with food stalls lining the streets in neighbourhoods like Yaowarat (Chinatown) and Sukhumvit, where vendors sell everything from fresh seafood to exotic fruits. Some of the most iconic street food dishes in Bangkok include pad Thai (stir-fried noodles with shrimp, tofu, and peanuts), som tam (green papaya salad with lime and chili), moo ping (grilled pork skewers), and mango sticky rice (a sweet dessert made with glutinous rice, mango, and coconut milk).

The street food vendors in Bangkok offer a mix of traditional Thai dishes and innovative twists on classic recipes. Vendors often prepare food right in front of customers, allowing diners to observe the cooking process and enjoy the fresh, vibrant flavours of Thai cuisine. The popularity of Bangkok's street food scene has made it a major attraction for tourists, who come from around the world to experience the city's authentic flavours. Bangkok's street food is emblematic of Thai culture, emphasizing freshness, flavour balance, and the communal experience of sharing meals.

Istanbul, Turkey

Istanbul's street food culture reflects its unique position as a bridge between Europe and Asia, blending flavours and techniques from both regions. Istanbul's bustling streets are home to a diverse array of street food, from traditional Turkish dishes to modern snacks inspired by global cuisine. Street food is deeply ingrained in Istanbul's daily life, with vendors selling food near busy areas like Taksim Square, the Grand Bazaar, and along the Bosphorus. Some iconic street foods in Istanbul include simit (a sesame-covered bread similar to a bagel), balık ekmek (grilled fish sandwiches typically served along the waterfront), kumpir (baked potatoes loaded with various toppings), and köfte (Turkish meatballs served with bread or rice).

Another popular street food in Istanbul is döner kebab, which consists of seasoned meat, typically lamb or chicken, cooked on a vertical rotisserie and sliced thinly to serve in pita or flatbread. Istanbul's döner vendors have elevated this simple dish into a beloved staple, showcasing the spices and flavours of Turkish cuisine. Street food in Istanbul is both convenient and deeply flavourful, offering locals and tourists alike a taste of the city's rich culinary heritage. The popularity of street food in Istanbul reflects Turkish values of hospitality, community, and an appreciation for fresh, bold flavours.

Mexico City, Mexico

Mexico City is renowned for its street food culture, which is deeply rooted in Mexican culinary traditions and the diverse flavours of its regions. Street food in Mexico City is a blend of Indigenous ingredients, Spanish influences, and local ingenuity, resulting in a vast array of dishes that are both delicious and accessible. The city's street food scene includes everything from simple snacks to hearty meals, with vendors selling food in neighbourhoods like Condesa, Coyoacán, and Centro Histórico. Some of the most iconic street foods in Mexico City include tacos al pastor (marinated pork tacos with pineapple), elote (grilled corn on the cob topped with mayonnaise, cheese, and chili powder), tortas (Mexican sandwiches with a variety of fillings), and tamales (corn dough stuffed with meat or cheese, wrapped in corn husks, and steamed).

Tacos, in particular, are central to Mexico City's street food culture. Vendors often prepare tacos on the spot, serving them with fresh toppings like cilantro, onions, and salsa. The variety of taco fillings, from carnitas (slow-cooked pork) to barbacoa (spiced lamb), showcases the diversity of Mexican flavours. Another popular street food is churros, a sweet snack made from fried dough and dusted with sugar, often served with chocolate sauce for dipping. Mexico City's street food reflects the warmth, vibrancy, and flavour complexity of Mexican culture, and it continues to be an essential part of daily life for residents and a major draw for food-loving travellers.

These cities—Bangkok, Istanbul, and Mexico City—each showcase the unique role of street food in urban culture, illustrating how food can shape a city's identity, provide affordable dining options, and bring people together. Each of these hubs offers a distinct culinary experience that reflects the cultural, historical, and social fabric of the city.

Practical Insights for Chefs on Integrating Street Food Trends into Menus

Street food's popularity and accessibility make it an appealing addition to modern restaurant menus. Chefs interested in integrating street food trends can bring the flavours, creativity, and communal spirit of street food to their kitchens, creating dishes that capture the excitement of this culinary tradition. Here are practical insights for chefs on incorporating street food trends into their menus with authenticity and respect.

1. Embrace Bold Flavours and Simplicity

Street food is often characterized by its bold flavours and straightforward preparation, making it ideal for chefs looking to add flavourful yet approachable dishes to their menus. Chefs can focus on creating dishes that showcase strong, well-balanced flavours with minimal ingredients, capturing the essence of street food. For example, a chef might create a simplified version of Thai pad kra pao (stir-fried basil chicken) by using fresh basil, garlic, chilies, and soy sauce, served over rice. Embracing simplicity allows the flavours to shine and makes the dish accessible to diners who may be unfamiliar with the cuisine.

2. Adapt Street Food for Restaurant Presentation

While street food is typically served on the go, chefs can adapt it for sit-down dining by enhancing presentation and portion sizes without losing authenticity. For example, instead of serving tacos in a traditional wrapper, chefs can present them on artisanal ceramic plates with fresh garnishes. The same flavours and ingredients can be elevated by thoughtful plating, adding a restaurant-level experience to a dish rooted in street culture. Adapting street food for restaurant settings allows diners to enjoy these flavours in a comfortable environment while still capturing the essence of the dish.

3. Incorporate Street Food as Small Plates or Shareable

Dishes

Street food's accessibility and variety make it ideal for small plates or shareable dishes on a restaurant menu. Chefs can create a selection of street food-inspired small plates, encouraging diners to sample a variety of flavours and dishes. For example, a chef might offer a mezze platter inspired by Middle Eastern street food, with items like falafel, hummus, tabbouleh, and pita bread. This approach allows diners to experience multiple dishes, fostering a communal dining experience that echoes the social nature of street food.

4. Use Seasonal and Local Ingredients

Many street food vendors rely on seasonal and local ingredients to keep their offerings fresh and affordable. Chefs can incorporate this practice by using locally sourced produce and ingredients when creating street food-inspired dishes. For instance, a chef might offer elote (Mexican-style grilled corn) when corn is in season, topping it with locally produced cheese or herbs. Using seasonal and local ingredients not only enhances the freshness of the dish but also supports local farmers and aligns with sustainable practices.

5. Educate Diners About the Cultural Significance of Street Food

Street food is often deeply connected to the culture and history of a region, and educating diners about this significance can enhance their appreciation of the dish. Chefs can provide brief descriptions on the menu, explaining the origins or traditional aspects of a dish, such as the history of Mexico's tacos al pastor or the popularity of Thai mango sticky rice in Bangkok's street food markets. This approach adds depth to the dining experience, fostering an appreciation for the cultural heritage behind each dish and allowing diners to connect with the food on a more meaningful level.

6. Collaborate with Street Food Vendors and Cultural Experts

For chefs who are not deeply familiar with a particular cuisine, collaborating with street food vendors or cultural experts can provide valuable insights and ensure authenticity. Working alongside a street food chef or vendor can help refine recipes, improve flavour profiles, and maintain the integrity of the dish. This collaboration is also a way to honour the source of inspiration and demonstrate respect for the traditions behind the food. Additionally, hosting guest chefs or creating limited-time pop-ups featuring street food experts can draw diners and offer an authentic street food experience.

7. Innovate with Fusion and Local Adaptations

Street food's adaptability makes it a natural fit for fusion, allowing chefs to experiment with flavours from different cultures. For instance, a chef might create a taco with Korean bulgogi as the filling, blending Mexican and Korean flavours in a way that honours both traditions. Alternatively, local ingredients can be incorporated into classic street food recipes, such as using a locally available fish for Japanese-style grilled skewers. Fusion dishes and local adaptations keep menus fresh and exciting while reflecting the global nature of modern street food culture.

Integrating street food trends into restaurant menus allows chefs to capture the excitement, bold flavours, and cultural significance of street food in a way that resonates with diners. By embracing simplicity, adapting presentation, using local ingredients, educating diners, collaborating with experts, and exploring fusion, chefs can create a dining experience that honours the heritage of street food while showcasing their unique culinary style.

CHAPTER 17: MODERN FOOD MOVEMENTS (E.G., FARM-TO-TABLE, ORGANIC)

Rise Of Modern Food Movements And Their Cultural Implications

In recent decades, the culinary world has witnessed the rise of various food movements that emphasize sustainability, ethical sourcing, health, and a return to natural and traditional agricultural practices. Movements like farm-to-table, organic, and locavore have gained widespread popularity, driven by a desire to create food systems that prioritize environmental health, animal welfare, and community well-being. These movements have changed the way people think about food, challenging industrial agriculture and advocating for practices that are both ecologically sustainable and socially responsible. As these movements have grown, they have had significant cultural implications, shaping the values and practices of chefs, farmers, and consumers alike.

The farm-to-table movement is one of the most influential modern food trends, emphasizing direct sourcing from local farms and a focus on fresh, seasonal ingredients. This movement emerged as a reaction against the rise of industrial agriculture and processed foods, which have dominated food systems since the mid-20th century. The farm-to-table philosophy advocates for shortening the distance between farms and consumers, ensuring that food reaches the table in its freshest and most nutritious form. For chefs, farm-to-table means building relationships with local farmers, supporting sustainable agriculture, and creating menus that change with the seasons. The movement promotes a sense of connection to the land and local community, celebrating the natural diversity of each region's produce.

The organic movement shares many values with farm-to-table but places a specific emphasis on avoiding synthetic chemicals, pesticides, and genetically modified organisms (GMOs) in food production. Organic farming practices are designed to support soil health, biodiversity, and water conservation, fostering an agricultural system that works in harmony with nature. Consumers have become increasingly aware of the potential health benefits of organic foods, as well as the environmental impact of pesticide use and synthetic fertilizers. As a result, organic foods have seen a surge in demand, with many restaurants and grocery stores offering organic options to meet this consumer interest. For chefs, organic ingredients provide an opportunity to showcase flavours that are pure, vibrant, and reflective of sustainable farming methods.

Another significant movement is the locavore movement, which encourages consumers to eat foods produced within a specific geographic area, typically within 100 miles. The locavore philosophy emphasizes reducing the environmental footprint associated with long-distance food transportation, supporting local economies, and eating foods that are in season. Locavores often prioritize food systems that reflect local culture and traditions, as well as the unique characteristics of regional produce. For chefs, embracing a locavore approach means designing menus around what's available nearby, adapting to seasonal changes, and finding creative ways to use the diverse ingredients produced in their local environment.

These modern food movements have cultural implications that go beyond what's on the plate. They represent a shift towards greater awareness of the impact of food choices on the environment, animal welfare, and human health. By embracing these movements, chefs, farmers, and consumers are contributing to a cultural landscape that values sustainability, respect for nature, and a closer connection to the sources of our food. As food movements continue to evolve, they encourage a holistic approach to dining that aligns with values of community, ethics, and environmental stewardship.

Case Studies of Successful Farm-to-Table Restaurants

The farm-to-table movement has inspired chefs worldwide to reimagine their approach to food sourcing and menu development. Successful farm-to-table restaurants have demonstrated that local, seasonal ingredients can create extraordinary dining experiences, each unique to its location and community. Below are a few examples of renowned farm-to-table restaurants that embody the values of this movement.

1. Chez Panisse (Berkeley, California)

Founded by Alice Waters in 1971, Chez Panisse is often credited with pioneering the farm-to-table movement in the United States. Waters' philosophy centres on sourcing ingredients directly from local farmers and showcasing the natural flavours of seasonal produce. Chez Panisse's menus change daily, reflecting the freshest ingredients available from nearby farms. The restaurant has long-standing relationships with local farmers, ranchers, and artisans, each committed to sustainable and organic practices. Waters' dedication to farm-to-table dining has made Chez Panisse a cultural landmark, inspiring a generation of chefs to focus on ingredient quality, simplicity, and ethical sourcing.

2. Blue Hill at Stone Barns (Pocantico Hills, New York)

Blue Hill at Stone Barns, led by chef Dan Barber, is located on a working farm and nonprofit agricultural centre dedicated to sustainable farming practices. Barber's approach to farm-to-table goes beyond sourcing; he integrates farming and cooking into a holistic experience. The restaurant's menus are entirely seasonal and feature ingredients grown and raised on-site, as well as from local suppliers committed to regenerative agriculture. Barber is known for his creative use of less conventional cuts of meat and vegetables, highlighting the importance of using the entire plant or animal. Blue Hill at Stone Barns is an exemplary model of a farm-to-table restaurant that prioritizes environmental sustainability, waste reduction, and a commitment to ethical food systems.

3. Narisawa (Tokyo, Japan)

Yoshihiro Narisawa's eponymous restaurant in Tokyo combines traditional Japanese techniques with a farm-to-table approach, sourcing ingredients from local producers who practice sustainable farming. Narisawa is known for his commitment to shun, the Japanese concept of seasonality, and he builds his menus around ingredients that are at their peak flavour each season. The restaurant's cuisine, often referred to as "innovative Satoyama," emphasizes the relationship between human life and nature, celebrating Japanese landscapes and ingredients through artful, environmentally conscious dishes. Narisawa's approach exemplifies how farm-to-table principles can be adapted to different cultural contexts, highlighting the beauty of local ingredients and sustainable practices in Japanese cuisine.

These case studies illustrate the success and creativity that can emerge from a commitment to farm-to-table values. Each of these restaurants showcases the unique flavours of its region, fosters relationships with local farmers, and demonstrates a dedication to sustainability. By focusing on seasonality, simplicity, and ethical sourcing, these restaurants serve as models for chefs looking to incorporate farm-to-table principles into their own kitchens.

Tips for Chefs on Ethical Sourcing and Sustainability in the Kitchen

As the demand for ethical sourcing and sustainability grows, chefs have an opportunity to align their practices with these values, creating a positive impact on the environment and their local communities. Integrating sustainable practices into the kitchen not only enhances food quality but also contributes to a broader movement toward responsible food systems. Here are practical tips for chefs looking to embrace ethical sourcing and sustainability.

1. Build Relationships with Local Farmers and Producers

One of the most effective ways to ensure ethical sourcing is to work directly with local farmers and producers. Building strong relationships with these suppliers allows chefs to understand where their ingredients come from, how they are grown or raised, and the practices used in their production. Chefs can visit farms, meet the people behind the produce, and select suppliers who prioritize organic, regenerative, or other sustainable practices. By purchasing locally, chefs support their community's economy and reduce the environmental impact of transporting food over long distances.

2. Prioritize Seasonal Ingredients

Using seasonal ingredients is fundamental to both farm-to-table and sustainable cooking practices. Seasonal produce is typically fresher, more flavourful, and requires fewer resources to grow compared to out-of-season ingredients. Chefs can work with their suppliers to determine which fruits, vegetables, and herbs are in season, incorporating these ingredients into their menus. Seasonal cooking not only reduces the kitchen's carbon footprint but also encourages creativity, as chefs experiment with the variety each season brings. This approach fosters a deeper appreciation for nature's cycles and challenges chefs to create dishes that reflect the natural rhythms of their environment.

3. Embrace Whole-Animal and Whole-Plant Cooking

Minimizing waste is an essential component of sustainability, and one way to achieve this is by practicing whole-animal and whole-plant cooking. Whole-animal cooking, for instance, encourages chefs to use every part of the animal, from the prime cuts to the offal and bones, creating diverse dishes and reducing waste. Similarly, whole-plant cooking involves using parts of vegetables that are often discarded, such as stems, leaves, and peels. This approach maximizes the yield of each ingredient and promotes respect for the resources that went into producing it. Chefs can incorporate these principles by creating dishes that utilize "secondary" cuts of meat or by making stocks, sauces, and garnishes from vegetable scraps.

4. Use Energy-Efficient Equipment and Reduce Water Waste

Sustainability in the kitchen goes beyond food sourcing; it also includes using energy and water efficiently. Chefs can adopt eco-friendly practices by using energy-efficient appliances, such as induction cooktops, LED lighting, and low-flow faucets. These changes reduce the kitchen's energy consumption and environmental impact. Additionally, chefs can implement water-saving techniques, such as reusing rinse water for cleaning or reducing water waste during food preparation. By making small adjustments in kitchen operations, chefs contribute to a more sustainable restaurant environment.

5. Offer Plant-Based Options

Offering plant-based dishes aligns with sustainable practices by reducing reliance on resource-intensive animal products. Plant-based foods have a lower environmental footprint, as they require less land, water, and energy to produce compared to meat and dairy. By incorporating plant-based options into their menus, chefs not only cater to a growing demand for vegetarian and vegan dishes but also contribute to reducing the kitchen's carbon footprint. Chefs can create inventive plant-based dishes that showcase seasonal vegetables, grains, and legumes, making sustainable dining exciting and accessible for all guests.

6. Educate Diners About Sustainable and Ethical Choices

Chefs have a unique opportunity to educate diners about the importance of ethical sourcing and sustainability. By sharing information about the origins of ingredients, the benefits of seasonal produce, or the farm-to-table philosophy, chefs can foster a deeper appreciation for sustainable food practices among their guests. Menus can include brief descriptions about the sourcing of key ingredients, or servers can provide information on the sustainability of certain dishes. This transparency builds trust and encourages diners to make informed choices, transforming the dining experience into a platform for promoting ethical and environmentally conscious food systems.

7. Practice Transparent Sourcing and Labelling

Transparency is key to maintaining the integrity of ethical sourcing. Chefs should be transparent about where their ingredients come from, especially when claiming that ingredients are organic, local, or sustainably sourced. By maintaining clear communication with suppliers and verifying the sustainability practices of each source, chefs can confidently communicate these values to their customers. Transparent sourcing not only supports sustainability but also builds a reputable brand that diners can trust.

8. Engage in Continuous Learning and Collaboration

The field of ethical sourcing and sustainability is constantly evolving, with new practices, technologies, and certifications emerging regularly. Chefs can stay informed by attending workshops, participating in sustainable food organizations, and collaborating with other chefs who share a commitment to ethical sourcing. By engaging in continuous learning, chefs can remain at the forefront of sustainable practices and adapt to new developments that enhance their kitchen's environmental and ethical impact.

In conclusion, embracing ethical sourcing and sustainability in the kitchen requires a commitment to responsible practices, from sourcing ingredients to minimizing waste and reducing energy use. By prioritizing local, seasonal ingredients, practicing whole-ingredient cooking, educating diners, and fostering transparency, chefs can play a significant role in creating a more sustainable and ethical food system. These practices not only enhance the dining experience but also reflect a respect for nature, local communities, and the future of food.

CHAPTER 18: FOOD AND TECHNOLOGY

Impact Of Technology On Food Preparation, Preservation, And Consumption

Technology has revolutionized the food industry in ways unimaginable just a few decades ago, impacting everything from how food is prepared and preserved to how it is consumed. The integration of technology in the kitchen has not only increased efficiency and safety but also introduced chefs to new culinary possibilities, enabling them to create innovative dishes and redefine traditional cooking techniques. Technological advancements have transformed both home and professional kitchens, allowing for more precise control over ingredients and cooking processes while expanding the scope of culinary creativity.

One of the earliest applications of technology in food preparation was the introduction of mechanical and electrical appliances, which simplified tasks that were once labour-intensive and time-consuming. From mixers and blenders to food processors and sous-vide machines, kitchen appliances have allowed chefs to prepare food with greater speed and consistency. For instance, the sous-vide technique, which involves vacuum-sealing ingredients and cooking them in a water bath at a controlled temperature, enables chefs to achieve a level of precision in cooking that would be difficult to replicate with traditional methods. This technique, made possible by advancements in temperature control and sealing equipment, allows for perfectly cooked proteins, vegetables, and even desserts, with flavours and textures that are retained and enhanced.

In food preservation, technology has played a crucial role in extending shelf life, reducing waste, and ensuring food safety. Refrigeration and freezing have been transformative in food storage, allowing perishable items to be kept for longer periods without losing quality. Additionally, techniques like freeze-drying, canning, and vacuum-sealing have made it possible to store and transport food across long distances, meeting the needs of global food systems. Freezing technology, in particular, has evolved to include methods like flash-freezing, which preserves the texture, flavour, and nutrients of food more effectively than traditional freezing. These advancements have not only improved access to a wider variety of foods year-round but have also helped reduce food waste by extending the life of seasonal produce.

Technology has also changed the way people consume food, particularly with the rise of online ordering platforms, meal delivery apps, and food-related social media. These platforms allow people to access restaurant-quality food from the comfort of their homes and have created new business opportunities for chefs and food entrepreneurs. Social media has transformed food from a purely sensory experience to a visually shareable one, influencing food trends and shaping culinary preferences on a global scale. Dishes that are visually appealing and innovative are more likely to go viral, encouraging chefs to think about presentation and design alongside flavour and technique. For chefs, social media serves as a platform to showcase their work, connect with a wider audience, and stay informed about emerging trends in the culinary world.

The impact of technology on food preparation, preservation, and consumption reflects a broader trend of innovation in the food industry. While technology has enhanced convenience and creativity in the kitchen, it has also raised questions about the potential loss of traditional practices and artisanal techniques. As chefs navigate this evolving landscape, they must consider how to balance the benefits of technology with the preservation of culinary heritage, ensuring that technology complements rather than replaces the skills and knowledge honed through years of culinary tradition.

Notable Advances Such as Molecular Gastronomy and Food Printing

In recent years, several groundbreaking technologies have emerged in the culinary world, redefining what is possible in the kitchen. Notable among these are molecular gastronomy and food printing, both of which have introduced entirely new approaches to cooking and presentation.

Molecular Gastronomy

Molecular gastronomy is a field of food science that explores the physical and chemical transformations of ingredients during cooking. This discipline, popularized by chefs like Ferran Adrià and Heston Blumenthal, has allowed chefs to push the boundaries of flavour, texture, and presentation by applying scientific principles to culinary techniques. Molecular gastronomy techniques often involve the use of unusual ingredients, such as liquid nitrogen, agar-agar, and sodium alginate, to create dishes that challenge traditional expectations of food.

One of the most famous techniques in molecular gastronomy is spherification, a method developed by Ferran Adrià at El Bulli. Spherification involves combining a liquid with sodium alginate, then dropping it into a bath of calcium chloride to create small, caviar-like spheres. These spheres, which can contain flavours as diverse as olive juice or fruit purée, burst in the mouth, creating an unexpected and memorable experience for diners. Another technique, foam creation, involves using a siphon to transform liquids into light, airy foams. Chefs use this method to create foams from ingredients like fruits, herbs, and even cocktails, adding a novel texture to dishes and enhancing the sensory experience.

Sous-vide cooking is also a staple of molecular gastronomy. By cooking ingredients in vacuum-sealed bags at low, precise temperatures, chefs can achieve consistent textures and retain the flavours of each ingredient. This method is especially popular for cooking proteins, as it allows for even cooking without overcooking, creating tender and juicy dishes. Sous-vide has become more accessible to home cooks due to advancements in affordable sous-vide machines, demonstrating how molecular gastronomy has influenced both professional and amateur kitchens.

Molecular gastronomy has not only expanded the possibilities of texture and presentation but has also opened new avenues for exploring flavour combinations. Techniques like gelification, dehydration, and carbonation have become part of the molecular chef's toolkit, each offering a unique way to transform ingredients and surprise diners. Molecular gastronomy encourages chefs to view cooking as a science, using their understanding of chemistry and physics to manipulate ingredients in unexpected ways.

Food Printing

One of the most recent technological innovations in the culinary world is food printing, which uses 3D printing technology to create intricate food designs. Food printers can produce precise shapes and textures by layering edible materials like chocolate, dough, or purées. This technology, still in its early stages, has the potential to revolutionize food presentation, allowing chefs to create designs and structures that would be nearly impossible to achieve by hand.

Food printing can be used to produce customized shapes, personalized messages, and elaborate designs, making it ideal for pastry work and dessert plating. For example, a food printer might be used to create a latticework of chocolate for a dessert or a perfectly shaped edible garnish. Food printing also has practical applications in terms of portion control and nutrient customization, allowing chefs to tailor dishes to meet specific dietary needs or preferences. As this technology advances, food printers may even be capable of creating more complex dishes with multiple layers of flavour and texture.

In addition to aesthetic applications, food printing is being explored as a solution for sustainable and alternative food sources. Researchers are experimenting with printed foods made from alternative proteins, such as algae, insects, and plant-based proteins, which can be incorporated into dishes with minimal environmental impact. This approach has the potential to make food production more efficient and sustainable, as it reduces waste and allows for precise control over ingredients.

While molecular gastronomy and food printing represent exciting advances in the culinary world, these techniques are not without controversy. Some critics argue that such technologies prioritize novelty over tradition and flavour, potentially alienating diners who prefer familiar, comforting food. However, when used thoughtfully, these innovations can enhance rather than replace traditional cooking methods, offering chefs new tools to elevate their cuisine and create memorable dining experiences.

Considerations for Chefs Using Technology Without Losing Tradition

As technology becomes increasingly prevalent in the culinary world, chefs face the challenge of integrating these innovations without losing touch with culinary tradition. Balancing modern techniques with respect for traditional methods is essential for chefs who wish to honour the authenticity and cultural heritage of their cuisine while exploring new possibilities in the kitchen. Here are some considerations for chefs using technology thoughtfully and respectfully.

1. Prioritize Flavour Over Novelty

In the pursuit of innovation, chefs may be tempted to prioritize novelty over taste. However, the most successful dishes are those that balance creativity with flavour. Technology can enhance flavour by allowing for precision and consistency, but it should not overshadow the primary goal of creating delicious food. Chefs can use techniques like sous-vide or spherification to enhance the natural flavours of ingredients, rather than relying on technology solely for its shock value. By focusing on taste as the central aspect of a dish, chefs can use technology as a tool to elevate rather than distract from the dining experience.

2. Use Technology to Enhance, Not Replace, Traditional Techniques

Many culinary traditions are rooted in techniques that have been passed down through generations, each with its unique cultural significance and flavour profile.

When incorporating technology, chefs can seek ways to enhance traditional methods rather than replace them entirely. For example, sous-vide can be used to enhance the tenderness of meats in a traditional stew, while maintaining the recipe's original spices and ingredients. In pastry work, 3D printing might be used to create intricate designs on a traditional cake or tart, preserving the original recipe while adding a modern touch. By combining technology with tradition, chefs create dishes that honour both the past and the present.

3. Consider the Cultural Context of Each Dish

Many traditional dishes carry cultural and historical significance, and altering them too dramatically can risk losing their essence. Chefs can be mindful of the cultural context of each dish they create, using technology in ways that respect the dish's origins. For example, when preparing a traditional Italian pasta dish, a chef might use a modern pasta extruder for consistency but adhere to traditional cooking techniques and ingredients. This approach ensures that the dish remains authentic while benefiting from technological advancements. Understanding the origins and meaning of a dish helps chefs avoid cultural insensitivity and create food that respects its heritage.

4. Educate Diners About the Technology Behind the Dish

Technology can add a layer of intrigue to the dining experience, and chefs can enhance this by educating diners about the techniques used to create each dish. Explaining how sous-vide cooking enhances flavour or describing the process of spherification can engage diners and help them appreciate the craft involved. This transparency not only demystifies technology but also reinforces the chef's commitment to both innovation and quality. Educating diners creates a more interactive dining experience, where guests feel connected to the creative process behind their meal.

5. Balance Innovation with Sustainability

As chefs incorporate technology into their kitchens, they should consider the environmental impact of their choices. Some advanced equipment, like 3D food printers or molecular gastronomy tools, can have a significant energy footprint. Chefs can strive to balance innovation with sustainability by choosing energy-efficient equipment, reducing waste, and using local or sustainably sourced ingredients whenever possible. Additionally, chefs can explore technology that supports sustainability, such as using vacuum-sealing for ingredient preservation or experimenting with alternative protein sources through food printing. By aligning technological advancements with sustainable practices, chefs contribute to a more responsible culinary industry.

6. Embrace Continuous Learning and collaboration

Technology in the culinary world is constantly evolving, and chefs can stay informed by engaging in continuous learning and collaborating with peers who specialize in different techniques. Attending workshops, experimenting with new tools, and learning from molecular gastronomy experts or food scientists can expand a chef's skill set and encourage responsible use of technology. By collaborating with other chefs and innovators, chefs gain insights into how technology can be applied thoughtfully and effectively, ensuring that their approach remains grounded in respect for both tradition and innovation.

The impact of technology on the culinary world is profound, offering chefs new ways to prepare, preserve, and present food. However, as chefs integrate modern techniques like molecular gastronomy and food printing into their kitchens, it is essential to prioritize flavour, respect tradition, and maintain a commitment to sustainability. By using technology as a complement to traditional methods, chefs can create dishes that honour culinary heritage while embracing the possibilities of modern innovation, enhancing the dining experience for today's discerning guests.

CHAPTER 19: FOOD AND GLOBALIZATION

How Globalization Has Influenced Modern Eating Habits

Globalization has reshaped the way people eat, cook, and perceive food, creating an interconnected culinary landscape where ingredients, techniques, and dishes move freely across borders. With advances in technology, transportation, and communication, food has become a powerful vehicle for cultural exchange, enabling flavours and traditions to transcend their places of origin and become part of the global culinary experience. This phenomenon has had a profound impact on modern eating habits, expanding dietary choices and exposing people to an unprecedented variety of cuisines.

One of the most visible effects of globalization is the widespread availability of international ingredients and dishes. In most urban centres today, supermarkets and grocery stores stock an array of ingredients from different parts of the world, making it possible to cook a Japanese meal in New York or prepare Italian pasta in Mumbai. Ingredients like soy sauce, quinoa, miso, and sriracha have become household staples in regions far from their origins, demonstrating the blending of food cultures that characterizes modern eating habits. This increased access has enabled home cooks and chefs alike to experiment with international flavours, fostering a culinary culture that values diversity and fusion.

Globalization has also influenced the rise of fusion cuisine, where chefs and home cooks combine elements from different culinary traditions to create something entirely new. This blending of flavours and techniques is especially popular in countries with multicultural populations, where dishes like sushi burritos, Korean tacos, and ramen burgers have gained popularity. Fusion cuisine reflects a modern desire for innovation and exploration, allowing people to enjoy flavours from multiple cuisines within a single meal. The popularity of fusion dishes illustrates how globalization has encouraged culinary creativity, pushing boundaries while highlighting the versatility of food as a medium for cultural expression.

In addition to expanding culinary options, globalization has led to the standardization and commodification of certain dishes, making them recognizable and accessible worldwide. Fast food chains like McDonald's, Starbucks, and KFC have spread globally, adapting their menus to local tastes while maintaining a recognizable brand identity. This has introduced certain globalized versions of American-style burgers, coffee, and fried chicken into diverse cultures, influencing eating habits and preferences, especially among younger generations. The presence of international fast-food chains in places like China, India, and Brazil has created a shared food culture that transcends national boundaries, making certain foods instantly recognizable across continents.

However, the globalization of food has also raised concerns about cultural homogenization and the erosion of local food traditions. As international dishes become mainstream, they often overshadow traditional foods, particularly in urban centres where convenience and novelty are prioritized. Younger generations, influenced by global food trends, may favour imported flavours and fast food over traditional dishes, leading to a gradual loss of culinary heritage. Furthermore, the demand for certain globalized foods has put pressure on local agriculture and resources, as regions adapt to produce ingredients for popular global dishes rather than for local consumption.

Globalization's influence on modern eating habits is therefore complex, with both positive and negative implications. While it has made culinary diversity more accessible and fostered a spirit of exploration, it has also created challenges for preserving local food traditions and ensuring that culinary heritage is not lost amid the appeal of novelty and convenience. For chefs, understanding the impact of globalization on food is essential, as it allows them to navigate the fine line between innovation and cultural authenticity, ensuring that the flavours and traditions of different cuisines are respected and celebrated.

Case Studies of "Globalized" Dishes (e.g., Sushi, Pizza, Tacos)

Several dishes have become global icons, transcending their countries of origin to become popular in nearly every corner of the world. These "globalized" dishes, such as sushi, pizza, and tacos, reflect the complex interplay between tradition, adaptation, and the impact of globalization on food culture. Each of these dishes tells a story of cultural exchange, adaptation, and the challenge of maintaining authenticity while catering to diverse tastes.

Sushi

Sushi, a Japanese dish that traditionally combines vinegared rice with seafood or vegetables, has become one of the most recognized and popular global foods. While sushi originated as a simple and elegant dish in Japan, focused on fresh, high-quality ingredients, it has been adapted in various ways to suit international palates. In the United States, for example, sushi is often made with ingredients like avocado, cream cheese, and spicy mayonnaise, which are not part of traditional Japanese sushi. The California roll, a popular American sushi creation, features imitation crab, avocado, and cucumber, wrapped in rice and seaweed, catering to the American preference for familiar flavours and ingredients.

The global popularity of sushi reflects both the appeal of Japanese aesthetics and the adaptability of sushi to different cultural preferences. However, this adaptation has raised questions about authenticity, as traditional Japanese sushi chefs often view these adaptations as diverging from the culinary discipline that defines true sushi-making. Despite this tension, sushi's journey from Japan to global fame illustrates how globalization allows traditional dishes to evolve, making them accessible to a broader audience while sparking conversations about the balance between authenticity and adaptation.

Pizza

Pizza, originally from Naples, Italy, is one of the most globalized foods, enjoyed and adapted in nearly every country. The traditional Neapolitan pizza, made with simple ingredients like tomatoes, mozzarella, olive oil, and basil, reflects the Italian philosophy of using fresh, high-quality ingredients. However, as pizza spread across the world, it evolved to suit local tastes and preferences. In the United States, pizza became thicker, cheesier, and often topped with an array of ingredients, from pepperoni to pineapple. In Japan, pizza is sometimes topped with unique ingredients like squid, mayonnaise, and corn, while in Brazil, toppings like heart of palm and green peas are popular.

Pizza's adaptability and versatility have made it a universal comfort food, but its transformation has sparked debates about authenticity. Traditional Italian pizzaiolos (pizza makers) often emphasize that true pizza is defined by specific ingredients and techniques, while global versions have taken creative liberties that diverge from Italian standards. The global appeal of pizza, however, reflects its ability to be both a canvas for culinary innovation and a symbol of cultural exchange, illustrating how traditional dishes can thrive in new forms when embraced by different cultures.

Tacos

Tacos, a staple of Mexican cuisine, have also become popular worldwide, with variations that reflect the flavours and preferences of different regions. Traditional tacos consist of a soft corn tortilla filled with meats like carnitas, al pastor, or barbacoa, and topped with onions, cilantro, and salsa. In the United States, however, tacos have been adapted to include hard-shell versions, as well as diverse fillings like fish, grilled vegetables, and even Korean-inspired bulgogi. The Korean taco, for example, blends Mexican and Korean flavours, using marinated beef, kimchi, and gochujang (Korean chili paste), wrapped in a tortilla.

The popularity of tacos reflects the versatility of the tortilla as a base and the adaptability of Mexican flavours to fusion cuisine. Mexican food purists may argue that these adaptations stray from the traditional roots of Mexican cuisine, which emphasizes simplicity and freshness. However, the spread of tacos and their fusion variations demonstrates how globalization can lead to exciting culinary hybrids that honour the spirit of the original while embracing new influences.

These case studies illustrate how certain dishes have been transformed through globalization, each balancing tradition with adaptation. While these globalized versions may differ from their origins, they showcase the dynamic nature of food, which can evolve and adapt while retaining its cultural essence.

Strategies for Chefs to Balance Authenticity with Globalization

For chefs, balancing authenticity with the influences of globalization presents both challenges and opportunities. In a world where diners are increasingly curious about international flavours, chefs have the unique responsibility of respecting cultural traditions while adapting to modern tastes. Here are some strategies for chefs seeking to navigate this balance thoughtfully.

1. Understand the Cultural Significance of Each Dish

Authenticity begins with an understanding of the cultural and historical context of a dish. Chefs can research the origins, traditional ingredients, and preparation methods associated with a dish before adapting it. For example, knowing that sushi-making in Japan emphasizes precise knife skills, high-quality ingredients, and minimalism can guide chefs in creating sushi-inspired dishes that honour these principles, even if adapted to local tastes. This respect for tradition enhances the dish's authenticity, allowing chefs to make informed choices when incorporating global flavours.

2. Adapt with Integrity and Purpose

When adapting traditional dishes to suit local preferences, chefs can aim to do so with integrity and purpose. Rather than changing ingredients solely for convenience or novelty, chefs can consider how their adaptations align with the dish's essence. For example, a chef adapting a classic Italian risotto with local vegetables might retain the technique of slow stirring and gradual stock addition, ensuring that the texture and richness remain true to Italian standards. This approach allows chefs to celebrate local flavours while preserving the core identity of the dish.

3. Highlight Regional Variations and Techniques

Many traditional dishes have regional variations within their countries of origin, offering chefs a range of authentic styles to explore. By highlighting these regional differences, chefs can introduce diners to the diversity within a single cuisine. For example, a chef might offer different types of tacos inspired by various regions of Mexico, from Yucatan's cochinita pibil to Baja-style fish tacos. This approach not only respects the dish's authenticity but also educates diners about the culinary richness of the region, showcasing the adaptability of traditional foods without altering their essence.

4. Create Fusion Dishes with a Respectful Approach

Fusion cuisine can be a powerful way to blend culinary traditions, but it requires a respectful approach to avoid cultural appropriation or dilution. Chefs can aim to balance flavours thoughtfully, combining elements from different cuisines in a way that highlights their compatibility rather than overwhelming each tradition. For instance, a chef creating a sushi taco might retain traditional Japanese flavours like wasabi and pickled ginger while incorporating Mexican influences like cilantro and salsa, ensuring that each component is distinct yet harmonious.

5. Educate Diners About the Origins and Adaptations of Dishes

One way to maintain authenticity in the context of globalization is to educate diners about the dish's origins and any adaptations made. By including descriptions on menus or sharing insights through servers, chefs can inform diners about the traditional roots of a dish and the inspiration behind any adaptations. For example, a chef might explain the origin of ramen and the specific reasons for adding local ingredients, helping diners appreciate both the authenticity and the adaptation. This transparency fosters a deeper connection to the food, enhancing diners' understanding and respect for the dish's cultural background.

6. Source Ingredients Ethically and Locally

Globalization has made it possible to source ingredients from around the world, but chefs can still prioritize ethical and local sourcing to support sustainability and authenticity. When preparing global dishes, chefs can look for local ingredients that mimic the texture or flavour of traditional components. For instance, using locally sourced fish instead of imported varieties in a sushi-inspired dish reduces environmental impact while ensuring freshness. Ethical sourcing also means supporting local producers, ensuring that globalization does not undermine local food economies and traditions.

7. Embrace Continuous Learning and Cultural Sensitivity

Globalization requires chefs to navigate cultural sensitivities carefully. Continuous learning, particularly from cultural experts or native chefs, can provide valuable insights into different culinary traditions. Chefs might take workshops, attend culinary exchanges, or collaborate with chefs from diverse backgrounds, deepening their understanding of the dishes they wish to adapt. This approach not only strengthens the chef's skills but also demonstrates respect for the cultural heritage of each cuisine, allowing for thoughtful adaptations that honour tradition.

Globalization has created a culinary world where chefs can explore international flavours, but this exploration requires a mindful approach to authenticity. By understanding cultural significance, adapting with integrity, highlighting regional diversity, and educating diners, chefs can balance the influences of globalization with respect for tradition. This thoughtful approach to globalized cuisine allows chefs to celebrate diversity while preserving the cultural richness that makes each dish unique.

CHAPTER 20: ETHICAL ISSUES IN MODERN GASTRONOMY

Overview Of Ethical Concerns In Food (E.g., Animal Rights, Fair Trade)

Modern gastronomy faces a range of ethical concerns that stem from the complex and often interconnected nature of global food systems. As chefs, consumers, and food industry professionals become more aware of the environmental and social impacts of their choices, issues like animal welfare, fair trade, labour rights, and environmental sustainability have taken centre stage. Addressing these ethical concerns is essential for creating a food industry that respects both people and the planet, fostering a culinary culture rooted in responsibility and empathy.

One of the most prominent ethical concerns in the food industry is animal welfare. The treatment of animals raised for food has sparked widespread debate, particularly as industrial farming practices— characterized by intensive confinement, lack of access to natural habitats, and rapid growth cycles—have become the norm. These practices raise serious ethical questions, as animals are often kept in overcrowded conditions with limited access to sunlight, space, or proper nutrition. The push for more humane and sustainable animal agriculture has led to alternative practices, such as pasture-raised, free-range, and organic farming, which aim to provide animals with better living conditions and reduce the environmental impact of meat production. For chefs, choosing ethically raised meat and dairy products can help promote a more humane approach to animal farming.

Another significant issue is fair trade and the ethics of labour in food production. Many of the ingredients used in modern gastronomy, such as coffee, chocolate, tea, and tropical fruits, are produced in developing countries, where workers often face poor wages, long hours, and hazardous working conditions. Fair trade certification aims to address these issues by ensuring that farmers and labourers receive fair compensation, safe working conditions, and access to resources that support sustainable farming practices. For chefs, sourcing fair trade ingredients is a way to support ethical labour practices and contribute to a more equitable food system, especially for marginalized communities involved in global agriculture.

Environmental sustainability is another ethical concern that intersects with issues of food production, animal welfare, and fair trade. Industrial agriculture, with its reliance on synthetic fertilizers, pesticides, and fossil fuels, contributes significantly to greenhouse gas emissions, water pollution, and soil degradation. Sustainable agriculture practices, including organic farming, regenerative agriculture, and permaculture, offer alternatives that aim to protect ecosystems, reduce carbon footprints, and preserve biodiversity. For chefs, choosing sustainably sourced ingredients, reducing food waste, and implementing energy-efficient practices are key ways to promote environmental stewardship in the kitchen.

The ethical concerns in modern gastronomy are complex and multifaceted, involving considerations of animal rights, labour conditions, environmental impact, and food justice. As chefs navigate these issues, they play a vital role in shaping the values and practices of the food industry, using their influence to advocate for ethical practices that support both human and ecological well-being.

Discussion of Controversial Practices and Sustainable Alternatives

The food industry has been criticized for several controversial practices that raise ethical questions about the treatment of animals, the exploitation of labour, and the environmental impact of food production. Understanding these practices and exploring sustainable alternatives is essential for chefs and food professionals who wish to adopt more responsible and humane approaches to their craft.

1. Factory Farming and Intensive Animal Agriculture

Factory farming, or intensive animal agriculture, is one of the most controversial practices in modern food production. In these systems, animals are raised in confined spaces with limited access to the outdoors, often in conditions that prioritize efficiency and cost over animal welfare. For example, broiler chickens are bred to grow quickly and are often kept in overcrowded barns, while dairy cows are confined to feedlots and subject to high-stress conditions. The ethical issues surrounding factory farming include the physical and psychological suffering of animals, the environmental impact of waste runoff, and the use of antibiotics to promote rapid growth, which contributes to antibiotic resistance.

Sustainable alternatives to factory farming focus on pasture-raised, free-range, and organic systems that prioritize animal welfare and environmental health. Pasture-raised animals have access to natural grazing areas, allowing them to exhibit natural behaviours and reducing the need for antibiotics. Free-range systems provide animals with more space and exposure to natural light, improving their quality of life. Organic farming prohibits synthetic pesticides, fertilizers, and growth hormones, promoting a more balanced ecosystem. Chefs can support these ethical practices by sourcing meat, dairy, and eggs from farms that adhere to humane and environmentally friendly standards, highlighting these choices on their menus to inform and educate diners.

2. The Exploitation of Labour in Food Production

The exploitation of labour in agriculture is a pervasive

issue, especially for crops like coffee, cocoa, and tea, which are often grown in regions with limited labour protections. Many workers in these industries are paid low wages, work in unsafe conditions, and lack access to health care and education. Child labour is also a concern, particularly in the cocoa industry, where children are often employed to perform physically demanding tasks in hazardous environments. These labour practices raise significant ethical questions about fairness, human rights, and the social responsibility of consumers and businesses.

Fair trade certification offers a sustainable alternative that seeks to protect workers' rights and improve their livelihoods. Fair trade organizations work directly with cooperatives, providing farmers with better prices, stable contracts, and access to resources for community development. In addition to fair trade, some companies are adopting direct trade models, where they build direct relationships with farmers to ensure fair compensation and sustainable farming practices. For chefs, choosing fair trade and direct trade products is a way to support ethical labour practices, contribute to social justice, and create a positive impact on the communities that produce their ingredients.

3. Monoculture and the Loss of Biodiversity

Monoculture, the practice of growing a single crop over large areas, is common in industrial agriculture and has serious implications for biodiversity and soil health. Crops like corn, soy, and wheat are often grown in monoculture systems, which deplete soil nutrients, increase vulnerability to pests, and require significant chemical inputs to maintain productivity.

This practice not only harms the environment but also reduces genetic diversity, making food systems more susceptible to disease and climate change.

Sustainable alternatives to monoculture include polyculture and regenerative agriculture, which emphasize crop diversity, soil health, and ecological balance. Polyculture involves growing multiple crops in the same space, which can improve soil fertility, reduce pest pressure, and enhance biodiversity. Regenerative agriculture goes further by incorporating techniques like cover cropping, crop rotation, and no-till farming to restore soil health and sequester carbon. Chefs can support these practices by sourcing ingredients from farms that prioritize biodiversity, highlighting these sustainable choices in their menus, and educating diners about the benefits of diverse and resilient food systems.

4. Overfishing and Unsustainable Seafood

Overfishing has led to the depletion of many fish species, threatening marine ecosystems and the livelihoods of fishing communities. Popular fish like tuna, cod, and salmon have been overharvested, leading to population declines and disruption of the ocean's food chain. Unsustainable fishing practices, such as bottom trawling and bycatch, further harm marine life by damaging habitats and unintentionally capturing non-target species.

Sustainable alternatives include responsible seafood sourcing, aquaculture, and sustainable fishing certifications. Organizations like the Marine Stewardship Council (MSC) and Aquaculture

Stewardship Council (ASC) certify fisheries and farms that adhere to sustainable practices, protecting fish populations and minimizing environmental impact. Chefs can contribute to sustainable seafood practices by choosing certified seafood, sourcing from local and responsible fisheries, and offering underutilized fish species to reduce pressure on popular stocks. By promoting sustainable seafood, chefs help protect marine biodiversity and ensure the long-term viability of ocean resources.

Guidelines for Chefs Committed to Ethical Culinary Practices

For chefs, committing to ethical culinary practices means making informed choices that prioritize animal welfare, environmental sustainability, and social responsibility. By adopting these guidelines, chefs can play a pivotal role in promoting ethical values within the food industry, offering diners meals that are not only delicious but also aligned with humane and sustainable principles.

1. Source Ingredients from Ethical and Transparent Suppliers

Ethical sourcing begins with transparency. Chefs should work with suppliers who can provide clear information about the origins, production methods, and labour practices associated with their products. When selecting meat, dairy, or eggs, chefs can prioritize suppliers who follow humane and sustainable farming practices, such as pasture-raised, free-range, or organic systems. For crops like coffee, cocoa, and tea, chefs can choose fair trade or direct trade options, ensuring that farmers are fairly compensated and work in safe conditions. Transparent sourcing builds trust with diners and allows chefs to confidently promote the ethical standards behind their ingredients.

2. Prioritize Seasonality and Local Sourcing

Using seasonal and locally sourced ingredients supports local farmers, reduces food miles, and allows chefs to create menus that reflect the natural rhythms of their region. Seasonal produce is often fresher, more flavourful, and less resource-intensive than imported, out-of-season options. By designing menus around what's available locally, chefs not only reduce their kitchen's environmental impact but also create a dining experience that is connected to the local landscape. Embracing seasonality fosters a deeper appreciation for natural cycles and reduces reliance on large-scale industrial agriculture.

3. Minimize Food Waste and Practice Nose-to-Tail and Root-to-Stem Cooking

Reducing food waste is essential for sustainable kitchen practices. Chefs can minimize waste by implementing nose-to-tail and root-to-stem cooking principles, which involve using every part of the animal or plant. For example, vegetable stems and leaves can be used in stocks or sauces, while less popular cuts of meat can be transformed into flavourful dishes. Additionally, chefs can find creative ways to repurpose leftovers, such as turning vegetable trimmings into soups or pickles. By maximizing the use of each ingredient, chefs reduce food waste, lower their kitchen's environmental footprint, and honour the resources that went into producing the food.

4. Educate and Engage Diners on Ethical Choices

Educating diners about ethical culinary practices enhances their dining experience and raises awareness of important issues in the food industry. Chefs can include information on menus about the sourcing of ingredients, the environmental benefits of sustainable practices, and the impact of fair trade or ethical sourcing. By engaging in conversations with diners, chefs create an interactive experience that fosters a greater understanding of the food's origins and encourages diners to make conscious choices. This educational approach transforms dining into a meaningful experience that supports ethical food practices.

5. Embrace Continuous Learning and Adaptation

The field of ethical culinary practices is constantly evolving, with new standards, certifications, and sustainable techniques emerging regularly. Chefs can stay informed by attending workshops, collaborating with other professionals, and participating in sustainable food organizations. Continuous learning enables chefs to adapt their practices to meet the highest ethical standards, ensuring that their commitment to sustainability and social responsibility remains relevant and impactful. By embracing an open mindset, chefs can lead by example, inspiring their teams and customers to prioritize ethical choices.

6. Balance Culinary Innovation with Ethical

Responsibility

While culinary creativity is essential to a chef's craft, it should be balanced with ethical responsibility. When experimenting with new ingredients or techniques, chefs can consider the environmental and social impact of their choices. For example, if a chef wishes to incorporate exotic ingredients, they might research the sourcing and environmental impact of those items, or consider using local alternatives. This balance ensures that culinary innovation does not come at the expense of ethical values, allowing chefs to create dishes that are both imaginative and responsible.

7. Support Food Justice Initiatives

Food justice seeks to address inequalities within the food system, ensuring that all communities have access to healthy, affordable, and culturally appropriate food. Chefs can support food justice by sourcing from local food cooperatives, working with farmers from underrepresented communities, and donating excess food to organizations that combat hunger. By aligning with food justice initiatives, chefs contribute to a more equitable food system, advocating for fair access and representation within the culinary world.

In conclusion, ethical culinary practices require a commitment to animal welfare, environmental sustainability, fair labour, and food justice. By sourcing responsibly, minimizing waste, educating diners, and continuously learning, chefs can make a positive impact on the food industry. These practices not only enhance the quality of the dining experience but also reflect a commitment to creating a more humane, sustainable, and just food system.

CHAPTER 21: THE ROLE OF FOOD IN NATIONAL IDENTITY AND DIPLOMACY

Exploration Of How Countries Use Food As A Symbol Of National Pride And Diplomacy

Food is one of the most powerful symbols of national identity, offering a tangible way for people to connect with their culture and heritage. As a daily necessity and a pleasure, food encapsulates a nation's history, geography, values, and social customs. For many countries, traditional dishes serve as a source of pride, representing the flavours, ingredients, and techniques that are unique to their culture. These dishes often become emblematic of national identity, fostering a sense of unity and belonging among citizens. Moreover, in a globalized world where cultural boundaries are increasingly blurred, food remains a key medium through which countries assert their distinct identities and showcase their heritage to the rest of the world.

The relationship between food and national pride is evident in celebrations, festivals, and traditions centred around local cuisine. For example, in Mexico, Día de los Muertos (Day of the Dead) celebrations include traditional foods like pan de muerto (a sweet bread shaped to resemble bones) and sugar skulls, which symbolize the connection between food, culture, and remembrance. In Japan, dishes like sushi and ramen are not only enjoyed domestically but have also become icons of Japanese culture internationally, representing the precision, aesthetic values, and respect for ingredients that define Japanese cuisine. In each of these cases, food serves as a bridge between past and present, allowing individuals to connect with their cultural roots while expressing national pride.

Beyond domestic pride, food also plays a significant role in cultural diplomacy, allowing nations to share their cuisine and values with the global community. Culinary diplomacy, or "gastrodiplomacy," is the practice of using food to foster goodwill and build positive relationships between countries. By sharing their cuisine with others, nations can promote cultural understanding, create a positive image abroad, and enhance their diplomatic efforts. For instance, Thailand's "Global Thai" program, initiated in the early 2000s, aimed to increase the number of Thai restaurants worldwide, promoting Thai culture and cuisine as part of a larger strategy to attract tourism and investment. By enjoying Thai food, diners around the world are introduced to Thai flavours, ingredients, and cooking techniques, fostering an appreciation for Thai culture and creating a sense of connection.

In many ways, culinary diplomacy is a form of soft power, where nations influence others through cultural appeal rather than force. Food has the ability to break down language barriers and bring people together, making it an effective tool for building relationships in both informal and formal diplomatic settings. State dinners, international food festivals, and culinary exchanges provide opportunities for countries to showcase their cuisine, enhancing cross-cultural understanding and promoting a positive image. For chefs, understanding the role of food in national identity and diplomacy provides insight into the ways cuisine can communicate cultural values and foster meaningful connections.

Case Studies of Culinary Diplomacy Initiatives (e.g., Korean Food Wave, French Culinary Diplomacy)

Several nations have successfully used culinary diplomacy to share their culture with the world, each tailoring their approach to reflect their unique culinary heritage. Case studies of initiatives like the Korean food wave and French culinary diplomacy highlight the ways in which food can serve as an ambassador for national identity.

Korean Food Wave (Hallyu)

South Korea's "Hallyu" wave, which refers to the global popularity of Korean culture, music, and entertainment, has also extended to Korean cuisine. The Korean government, recognizing the international appeal of Korean pop culture, has used this momentum to promote Korean food as part of a broader cultural diplomacy strategy. Through initiatives like the "K-Food" campaign, the government has worked to introduce Korean ingredients, dishes, and culinary traditions to new audiences, emphasizing the health benefits and unique flavours of Korean cuisine. Signature dishes like kimchi, bibimbap, and bulgogi have become internationally recognized, each offering a taste of Korea's culinary heritage.

The Korean food wave is supported by a strong presence of Korean restaurants worldwide, as well as media exposure through popular Korean television shows, dramas, and cooking programs. Through platforms like these, international audiences are introduced to the art of Korean barbecue, the tradition of communal dining, and the health-promoting aspects of fermented foods like kimchi. By framing Korean cuisine as both flavourful and nutritious, South Korea has successfully positioned its food as a desirable aspect of modern, global dining culture. This initiative has not only boosted South Korea's cultural visibility but has also strengthened its economic ties with other countries, as Korean food products become more widely exported and available.

French Culinary Diplomacy

France, long celebrated for its culinary heritage, has historically used food as a symbol of national prestige and diplomacy. French cuisine is known for its emphasis on technique, refinement, and the art of dining, values that align with France's image as a cultural and intellectual centre. Recognizing the role of cuisine in its national identity, the French government has actively promoted French culinary traditions through various diplomatic channels, including state dinners, culinary schools, and international food festivals. French culinary diplomacy is grounded in the belief that food is an essential part of culture, deserving of recognition and preservation.

In 2010, UNESCO recognized French gastronomy as part of the Intangible Cultural Heritage of Humanity, underscoring the significance of French culinary practices. This acknowledgment further solidified France's role as a leader in the global culinary community. The French Ministry of Foreign Affairs has also organized events like "Goût de France" (Good France), an annual celebration where chefs worldwide prepare French-inspired meals to promote French cuisine and foster cultural exchange. By sharing French culinary techniques and values, this initiative encourages a global appreciation for the craft of French cooking, elevating France's reputation as a culinary destination.

These case studies illustrate how culinary diplomacy can strengthen national identity, foster international goodwill, and promote cultural appreciation. Both South Korea and France have leveraged their culinary heritage to create positive images on the world stage, using food as a medium for storytelling, cultural preservation, and relationship building.

Practical Advice for Chefs on Incorporating National Identity into Cuisine Respectfully

For chefs, incorporating elements of national identity into their cuisine offers an opportunity to celebrate cultural heritage and create meaningful connections with diners. However, this requires a thoughtful and respectful approach that balances authenticity with creativity. Below are practical guidelines for chefs who wish to infuse their cuisine with elements of national identity.

1. Honour Traditional Techniques and Ingredients

One of the most effective ways to incorporate national identity into cuisine is by honouring traditional techniques and ingredients. Chefs can research the methods used in traditional dishes, from cooking techniques to plating styles, and incorporate these elements into their own creations. For instance, a chef preparing Mexican-inspired cuisine might focus on traditional techniques like nixtamalization (a process used to prepare corn for masa) or the use of a comal (a flat griddle) for cooking tortillas. By preserving the techniques that define a cuisine, chefs pay homage to the cultural knowledge and culinary heritage that underpin each dish.

Using authentic ingredients is equally important, as certain flavours and textures are essential to maintaining the integrity of a dish. Chefs can source key ingredients from reputable suppliers or find local alternatives that closely resemble the original flavours. For instance, if sourcing traditional Japanese ingredients, chefs might seek out suppliers who specialize in high-quality miso, soy sauce, and dashi, ensuring that their dishes capture the depth of Japanese umami flavours. By respecting the core elements of traditional cuisine, chefs create a foundation of authenticity that allows them to celebrate national identity with integrity.

2. Understand the Cultural Significance of Dishes

Before adapting or modernizing traditional dishes, chefs should strive to understand the cultural significance behind each recipe. Certain dishes are deeply tied to rituals, celebrations, or historical events, making them more than just a collection of ingredients. For example, baklava holds special significance in Middle Eastern cultures, often associated with religious festivals and family gatherings. Similarly, dishes like paella in Spain or dim sum in China carry regional and historical meanings, each reflecting the values and social customs of their place of origin.

Understanding these cultural connections helps chefs make informed choices when adapting traditional dishes, ensuring that their interpretations are respectful and well-considered. By recognizing the importance of a dish within its cultural context, chefs can avoid unintentionally diminishing or misrepresenting its significance, allowing diners to experience the cuisine in a way that honours its origins.

3. Use Storytelling to Connect Diners with the Dish's Heritage

Storytelling is a powerful tool for conveying the cultural and historical significance of a dish. By sharing the story behind each dish, chefs create a deeper connection between the food and the diner, transforming the meal into a meaningful cultural experience. Menus can include brief descriptions that explain the dish's origins, regional variations, or traditional methods, helping diners appreciate the heritage behind each plate.

For instance, a chef serving Italian-inspired dishes might explain the origins of risotto in Northern Italy, where the climate is ideal for rice cultivation, or share the history of pesto in Liguria. This approach not only enriches the dining experience but also fosters cultural appreciation, encouraging diners to view each dish as a reflection of the people and places that created it. By using storytelling as a bridge, chefs build an atmosphere of respect and curiosity around the food, allowing national identity to shine through.

4. Innovate Mindfully and with Purpose

Innovation is a valuable part of culinary expression, but when working with traditional dishes, it should be approached mindfully. Chefs can innovate while respecting national identity by making thoughtful adaptations that enhance rather than overshadow the dish's original elements. For example, a chef might use a traditional French technique like sous-vide to prepare an Indian-inspired dish, allowing the flavours of Indian spices to blend while maintaining the texture and moisture of the ingredients.

Chefs can also experiment with presentation, offering a modern twist that preserves the flavours and essence of the dish. For instance, serving ramen in small, individual bowls as part of a tasting menu allows for creative presentation while honouring the flavours and textures that define Japanese ramen. Mindful innovation ensures that national identity is preserved, while allowing chefs to explore new interpretations that resonate with contemporary audiences.

5. Collaborate with Cultural Experts and Culinary Historians

When incorporating elements of a cuisine with which they are less familiar, chefs can benefit from collaborating with cultural experts, culinary historians, or chefs who specialize in that cuisine. Working with individuals who possess a deep understanding of the cultural and historical aspects of a cuisine provides valuable insights that enhance authenticity and respect. For example, a chef interested in adding Peruvian elements to their menu might consult a Peruvian chef or culinary historian to learn about traditional flavours, ingredients, and techniques. This collaboration enriches the chef's understanding and allows them to represent the cuisine with authenticity.

6. Encourage Cross-Cultural Dialogue in the Kitchen

Culinary exchange within a kitchen fosters cross-cultural dialogue, allowing chefs to learn from each other's traditions and techniques. By inviting team members from diverse backgrounds to share their culinary heritage, chefs can incorporate authentic elements from different cultures while respecting the knowledge and experiences of those who represent them. This approach creates a collaborative and respectful environment, ensuring that national identity is celebrated with authenticity and that the menu reflects the diversity and unity of the culinary team.

Incorporating national identity into cuisine requires a thoughtful approach that balances authenticity with creativity. By honouring traditional techniques, understanding cultural significance, using storytelling, and collaborating with experts, chefs can create dishes that celebrate national pride and foster cross-cultural understanding. This respectful approach allows chefs to use their craft as a means of cultural diplomacy, building connections that extend beyond the plate.

CHAPTER 22: FOOD AND GENDER ROLES

Historical Roles Of Men And Women In Food Production And Culinary Practices

Throughout history, food production, preparation, and consumption have often been influenced by gender roles, with distinct expectations placed upon men and women regarding their contributions to the culinary sphere. These roles, shaped by social, cultural, and economic factors, have varied across time and geography, with men and women frequently occupying different spaces within the culinary world. While these roles have evolved over time, the legacy of gendered divisions in food production and preparation continues to influence contemporary culinary practices and perceptions of male and female chefs.

Historically, women have been primarily responsible for domestic cooking and household food preparation. In many societies, women were expected to oversee the kitchen, feeding their families, preparing meals, and managing household provisions. The act of cooking was often viewed as a nurturing role tied to motherhood, domesticity, and care for the family. In agrarian societies, women's roles extended to the processing and preservation of food, as well as managing small-scale livestock and kitchen gardens that provided additional sustenance. Despite the essential nature of these tasks, women's contributions to food production were often undervalued, as they were seen as extensions of domestic responsibility rather than skilled labour.

Men, on the other hand, were traditionally more involved in hunting, fishing, and large-scale agricultural tasks, which required physical strength and often took place outside the domestic sphere. In many hunter-gatherer societies, men were responsible for hunting animals, while women gathered fruits, vegetables, and edible plants. This division was partly based on physical demands but also rooted in social structures that positioned men as providers. In agricultural societies, men typically worked in the fields, ploughing and cultivating staple crops, while women focused on food preservation, dairy production, and other labour-intensive activities that could be done closer to the home.

Interestingly, in the realm of professional cooking, men have traditionally held more prominent roles, particularly in high-status culinary positions. Historically, professional kitchens were male-dominated spaces, with men recognized as "chefs" or "cooks" in restaurants, royal courts, and institutions. This trend is still evident in many parts of the world, where top culinary positions are often held by men, despite women's longstanding involvement in cooking within the domestic sphere. The French culinary tradition, for instance, has long celebrated the "chef" as a male figure, associated with technical skill, creativity, and innovation. Women's cooking, in contrast, was often viewed as simpler and less prestigious, relegated to the domestic sphere rather than the professional arena.

While these gendered roles in food production have evolved significantly, they continue to shape perceptions and expectations within the culinary world. For chefs today, understanding the historical context of gendered food roles offers insights into the ways these divisions have influenced contemporary kitchens, as well as the importance of fostering inclusivity and challenging stereotypes in the industry.

Case Studies of Gendered Food Roles Across Various Cultures (e.g., Japanese Tea Ceremony, Italian Kitchens)

Gendered roles in culinary practices are evident across various cultures, each offering unique insights into how societies have assigned specific responsibilities to men and women within the realm of food. Examining these roles provides a deeper understanding of the cultural significance behind gendered culinary practices and the values associated with them.

Japanese Tea Ceremony (Chaji)

The Japanese tea ceremony, or chaji, is a traditional ritual that embodies Japanese aesthetics, spirituality, and hospitality. Historically, the tea ceremony was performed by both men and women, although their roles and experiences within this practice varied. Initially, tea was introduced to Japan by Buddhist monks, and the ceremony was practiced predominantly by men within Zen temples. Over time, tea culture evolved, becoming a ritual practiced within aristocratic circles and later embraced by samurai, artists, and merchants. Notable tea masters, such as Sen no Rikyū, formalized the principles of the tea ceremony, emphasizing harmony, respect, purity, and tranquillity.

In modern Japan, the tea ceremony is practiced by both men and women, but women tend to play a more significant role in its instruction and preservation. Women are often seen as guardians of traditional tea practices, teaching the ceremony and performing it within households, cultural centres, and schools. However, the tea ceremony remains a highly gendered practice, with women expected to embody certain ideals, such as grace, humility, and poise, while performing it. Male tea masters, on the other hand, are often viewed as authorities on tea culture and are more likely to be involved in innovative or avant-garde interpretations of the ceremony. This division reflects the broader gender expectations within Japanese culture, where women's roles are often associated with tradition and refinement, while men's roles are linked to innovation and leadership.

Italian Kitchens and the Role of Nonna

In Italy, the kitchen holds a special place within family life, often serving as the heart of the home and a space where culinary traditions are passed down through generations. In many Italian households, the role of the nonna (grandmother) is central to the preservation of regional recipes, techniques, and family culinary secrets. Nonna is often responsible for making dishes like pasta, sauces, and bread from scratch, using recipes and techniques that have been passed down orally. This role is more than just cooking; it represents a connection to Italian heritage, as grandmothers are seen as custodians of family and regional identity.

While nonna's influence is celebrated within the home, Italian professional kitchens have traditionally been dominated by men, with women often excluded from the high-status roles of chefs and sous-chefs. This division has created a dichotomy where women are seen as nurturers within the home kitchen, responsible for passing on traditions, while men are recognized as professionals and innovators within the restaurant industry. However, recent shifts in Italy's culinary world have seen more women breaking into professional kitchens, challenging the notion that the culinary arts are a male-dominated domain. This shift represents a growing recognition of women's contributions to Italian cuisine, both within the home and in the public sphere.

French Haute Cuisine and the Male Chef Persona

France's culinary tradition has historically celebrated the male chef as the epitome of skill, creativity, and artistry. French haute cuisine, a refined style of cooking that emphasizes technical mastery and innovation, has been heavily influenced by male chefs like Auguste Escoffier and Paul Bocuse, who defined French cooking standards. This perception has contributed to the male-dominated culture within French professional kitchens, where men have traditionally occupied prestigious roles, while women were often relegated to supporting positions or worked in domestic kitchens.

In recent years, however, female chefs in France have gained greater recognition, challenging the stereotype that only men can master the art of haute cuisine. Notable female chefs, such as Anne-Sophie Pic and Dominique Crenn, have broken through the glass ceiling of French cuisine, earning Michelin stars and acclaim for their culinary prowess. Despite these advancements, the lingering perception of the "male chef" persona in French haute cuisine illustrates the historical gender roles that continue to shape professional kitchens. By celebrating female chefs and their contributions, the French culinary world is gradually embracing a more inclusive vision of haute cuisine, where skill and creativity are valued regardless of gender.

These case studies highlight the ways in which gendered roles in food production and culinary practices have shaped perceptions of men and women in different cultures. For chefs, understanding these cultural nuances offers valuable insights into the ways gender expectations have influenced the culinary world, underscoring the importance of challenging stereotypes and fostering inclusivity in contemporary kitchens.

Insights for Chefs on Breaking Stereotypes and Promoting Inclusivity in the Culinary World

In today's culinary landscape, chefs have the opportunity to challenge traditional gender roles and create inclusive environments that celebrate diversity and equality. By fostering a culture of inclusivity, chefs can empower team members of all genders to excel in the kitchen, promoting a culinary industry that values talent and creativity over stereotypes. Here are some strategies for chefs who wish to break gender stereotypes and promote inclusivity in the culinary world.

1. Recognize and Challenge Gender Stereotypes in the Kitchen

One of the first steps toward promoting inclusivity is recognizing and challenging the stereotypes that exist within the kitchen. Traditional gender roles often dictate which tasks are assigned to men and women, with men expected to handle heavy lifting, high-pressure tasks, or meat preparation, while women are assigned to pastry, garnishes, or support roles. Chefs can challenge these assumptions by assigning tasks based on skill and interest rather than gender, giving all team members the opportunity to work in different areas of the kitchen. This approach not only promotes skill development but also breaks down barriers, encouraging a more diverse and dynamic team.

2. Promote Women and Underrepresented Groups to Leadership Roles

Leadership roles in professional kitchens have traditionally been male-dominated, but chefs can help change this by promoting talented women and underrepresented individuals to positions of authority. By creating mentorship opportunities and providing support, chefs can empower women and minorities to take on leadership roles, showcasing their skills and perspectives. Promoting diversity in leadership fosters a more inclusive environment and challenges the perception that high-status culinary roles are reserved for men. This approach also provides role models for aspiring chefs from diverse backgrounds, demonstrating that talent and dedication are the primary qualifications for success.

3. Create an Inclusive and Supportive Kitchen Culture

Inclusivity goes beyond hiring practices; it requires creating a kitchen culture where all team members feel respected, valued, and supported. Chefs can set the tone by fostering open communication, encouraging collaboration, and addressing any behaviour that undermines inclusivity. Inclusive language, equitable policies, and a zero-tolerance approach to discrimination or harassment are essential components of a supportive kitchen environment. By prioritizing respect and understanding, chefs can create a culture where team members feel comfortable expressing themselves, sharing their ideas, and pursuing growth without fear of bias.

4. Celebrate Diverse Culinary Perspectives

Promoting inclusivity also involves celebrating the diverse culinary perspectives that each team member brings to the kitchen. Chefs can encourage team members to share their cultural backgrounds, culinary traditions, and unique skills, enriching the kitchen with a variety of flavours, techniques, and ideas. This approach not only enhances the creativity and authenticity of the menu but also fosters a sense of belonging among team members. By valuing diverse perspectives, chefs create a kitchen that reflects the diversity of the culinary world, allowing each team member to contribute meaningfully to the overall vision.

5. Support Work-Life Balance for All Team Members

In many kitchens, the expectation of long hours and intense workloads can create challenges for team members with caregiving responsibilities, a role traditionally associated with women. Chefs can promote inclusivity by supporting work-life balance for all team members, implementing flexible schedules, and offering support for those with family responsibilities. By acknowledging the importance of work-life balance, chefs create an environment where team members can thrive both personally and professionally, allowing them to pursue a fulfilling culinary career without sacrificing other aspects of their lives.

6. Address Pay Equity and Fair Compensation

Pay equity is a crucial aspect of promoting inclusivity in the culinary world. Historically, women and underrepresented groups have faced pay disparities, even in the same roles as their male counterparts. Chefs can address this issue by ensuring fair and transparent compensation practices, conducting regular pay reviews, and advocating for equal pay for equal work. By prioritizing pay equity, chefs contribute to a more just and inclusive industry where team members are compensated based on their skills and contributions, not their gender.

7. Highlight and Celebrate Female and Minority Culinary Achievements

Celebrating the achievements of female and minority chefs helps challenge stereotypes and inspire future generations of chefs. Chefs can highlight the accomplishments of women and underrepresented individuals within their teams, as well as showcase the contributions of notable figures in the culinary world. By acknowledging these achievements, chefs create a culture of appreciation and respect, empowering team members from diverse backgrounds to pursue excellence and make their mark on the culinary industry.

Breaking gender stereotypes and promoting inclusivity in the culinary world requires a commitment to equity, respect, and openness. By challenging traditional roles, supporting diversity in leadership, and fostering an inclusive kitchen culture, chefs can create an environment where talent, creativity, and passion are valued above all else. This approach enhances the team's dynamic and reflects modern gastronomy's evolving values, where diversity and inclusivity are celebrated as essential components of a vibrant and innovative culinary industry.

CHAPTER 23: FESTIVALS AND FEASTS: SEASONAL FOOD CELEBRATIONS

The Significance Of Food In Cultural Festivals And Seasonal Celebrations Worldwide

Throughout history, food has held a central place in festivals, feasts, and seasonal celebrations across diverse cultures. These gatherings often revolve around meals that highlight the unique flavours, ingredients, and culinary traditions of a region, creating a shared experience that brings families and communities together. Food is more than just sustenance during these occasions; it is a symbol of cultural identity, a means of honouring the harvest, and a way of expressing gratitude, joy, and togetherness. Seasonal celebrations allow people to connect with the cycles of nature, marking the passage of time through culinary traditions that resonate deeply within their cultural heritage.

Seasonal celebrations and festivals often coincide with the agricultural calendar, celebrating moments of abundance and preparing for times of scarcity. Harvest festivals, for instance, are common across many cultures and are a time to give thanks for the bounty of the land. Celebrations like Thanksgiving in the United States, Pongal in South India, and the Mid-Autumn Festival in China all honour the earth's produce, expressing gratitude for a successful harvest. These festivals feature meals that incorporate locally grown ingredients, from pumpkins and maize in North America to mooncakes and taro in East Asia. By using seasonal ingredients, these festivals celebrate the land's bounty and foster a deep connection to nature.

Religious festivals also play a significant role in food traditions, with each faith incorporating unique dishes and customs into its celebrations. During Diwali, the Hindu festival of lights, families prepare sweets like laddoos and barfis as offerings to the gods and as a symbol of good fortune. In Islam, the festival of Eid is marked by feasts that often include dishes like biryani, kebabs, and sweets to break the fasting month of Ramadan. For Christmas, Christians around the world celebrate with a range of dishes, from roast meats in Western countries to tamales and pozole in Latin America. These religious feasts serve as an expression of devotion, offering a way for individuals to observe their faith through shared meals that bring families and communities together.

In addition to religious and harvest festivals, seasonal celebrations mark key points in the calendar, with food playing a central role. In Japan, Hanami, or cherry blossom viewing, takes place in spring, where families and friends gather under blooming cherry trees to enjoy picnics featuring seasonal dishes like sakura mochi (a sweet rice cake with cherry blossom flavour) and bentos packed with colourful spring ingredients. Similarly, the Midsummer festival in Scandinavia celebrates the arrival of summer with outdoor feasts that feature fresh fish, new potatoes, strawberries, and local herbs. These celebrations reflect a reverence for seasonal change, emphasizing fresh ingredients that align with the natural cycles of growth and renewal.

Food's significance in cultural festivals lies not only in its flavours but also in the rituals, meanings, and memories attached to it. Each dish tells a story of cultural values, agricultural practices, and family traditions, fostering a sense of belonging and continuity. By partaking in these celebrations, individuals connect with their heritage, strengthen bonds with family and friends, and preserve culinary traditions for future generations. For chefs, understanding the role of food in seasonal and cultural festivals provides insight into the deep cultural resonance of certain dishes and ingredients, allowing them to incorporate elements of these traditions into their own culinary creations.

Case Studies of Traditional Feasts and Their Symbolic Dishes (e.g., Thanksgiving, Diwali, Lunar New Year)

Several festivals around the world are celebrated with traditional feasts that feature dishes rich in symbolism, reflecting the values and customs of the culture. Below are a few notable examples of such feasts and the foods that hold special meaning within them.

Thanksgiving (United States)

Thanksgiving is a North American harvest festival celebrated primarily in the United States and Canada, marking a time to give thanks for the year's harvest and express gratitude for family and friends. The centrepiece of a traditional Thanksgiving meal is the roast turkey, which symbolizes abundance and sustenance. Side dishes like cranberry sauce, pumpkin pie, stuffing, and sweet potatoes are also integral to the meal, each reflecting the autumnal ingredients that are locally harvested during this season. Cranberries, for instance, are a native North American fruit, while pumpkin, corn, and potatoes highlight the staples grown by Indigenous communities.

Thanksgiving is a deeply symbolic meal, reflecting the cultural values of gratitude, community, and sharing. The act of gathering around a large table and sharing dishes prepared with care is central to the celebration. The holiday's roots are associated with the Pilgrims' 1621 feast with the Wampanoag people, symbolizing cooperation and unity. For many families, Thanksgiving also includes personal traditions and recipes passed down through generations, making it a meaningful occasion that combines collective and individual cultural heritage.

Diwali (India)

Diwali, the Hindu festival of lights, is one of India's most important celebrations, symbolizing the triumph of light over darkness and good over evil. Food plays a central role in Diwali, with sweets, or mithai, holding particular significance. Families prepare and share sweets like laddoos (sweet balls made of gram flour and sugar), gulab jamun (fried milk dumplings soaked in syrup), and barfis (dense milk-based sweets flavoured with nuts or cardamom). These sweets symbolize prosperity, joy, and the sharing of good fortune, and are often offered to deities as part of the rituals.

In addition to sweets, savoury snacks like samosas, kachoris, and chaats are commonly enjoyed during Diwali. The diversity of dishes reflects India's culinary richness, with each region contributing unique flavours and recipes to the celebration. Diwali meals are often vegetarian, emphasizing purity and spirituality, and are enjoyed in the company of family and friends. For many, the preparation and sharing of food during Diwali embodies the spirit of generosity, unity, and blessings for the year ahead.

Lunar New Year (China and East Asia)

The Lunar New Year, celebrated across China and other East Asian countries, marks the beginning of the lunar calendar and is a time to honour ancestors, welcome new beginnings, and bring good fortune. Traditional foods enjoyed during this celebration are rich in symbolism, with each dish believed to bring specific blessings for the coming year. Dumplings are one of the most popular dishes, as their shape resembles ancient Chinese currency, symbolizing wealth and prosperity. Fish, often served whole, represents abundance and unity, as the Chinese word for fish (鱼, yú) sounds like the word for surplus.

Other symbolic dishes include niangao (sticky rice cake), which represents progress and success, and spring rolls, which are associated with wealth due to their gold-bar-like shape. The significance of each dish goes beyond its taste; it reflects a cultural belief in the power of food to influence one's fortune and future. During the Lunar New Year, families gather for a reunion dinner, a deeply meaningful meal that strengthens family bonds and upholds traditions that have been passed down through generations.

These case studies highlight the symbolic meanings attached to food in cultural festivals, illustrating how dishes can convey values, hopes, and cultural identity. For chefs, these celebrations offer inspiration to explore the symbolic aspects of ingredients and dishes, allowing them to create menus that honour the cultural heritage behind these feasts.

Creative Ways for Chefs to Incorporate Seasonal and Festive Elements into Their Culinary Offerings

For chefs, seasonal celebrations and festivals offer a wealth of inspiration to incorporate festive elements into their culinary offerings. By embracing the ingredients, flavours, and symbolism associated with various cultural festivals, chefs can create unique dining experiences that connect diners with the joy and richness of seasonal and cultural traditions. Below are creative strategies for chefs to infuse seasonal and festive elements into their menus.

1. Highlight Seasonal Ingredients with a Festive Twist

One of the simplest ways to incorporate seasonal and festive elements is by focusing on ingredients that are abundant and meaningful during specific times of the year. Chefs can design dishes around ingredients like pumpkins and cranberries in autumn, fresh herbs and greens in spring, or citrus fruits in winter. These ingredients not only reflect the season's offerings but also evoke the flavours associated with certain holidays and celebrations. For example, a chef might create a roasted pumpkin risotto with sage and nutmeg for a fall-inspired dish, capturing the essence of harvest festivals.

By emphasizing seasonal ingredients, chefs can create menus that are both sustainable and reflective of the natural cycles that many festivals celebrate. This approach also allows diners to enjoy flavours that are tied to the season, fostering a deeper connection to the rhythms of nature.

2. Offer Festive Tasting Menus Inspired by Global Festivals

Chefs can create tasting menus that draw inspiration from various cultural festivals, introducing diners to a range of traditional flavours and festive dishes. For instance, a Diwali-inspired tasting menu could feature samosas, spiced curries, and mithai-inspired desserts, showcasing the vibrant flavours and rich textures associated with the festival. Similarly, a Lunar New Year menu could incorporate dumplings, steamed fish, and sticky rice cake, each dish reflecting the symbolism of good fortune and unity.

By offering tasting menus that celebrate global festivals, chefs provide diners with an immersive cultural experience. These menus not only introduce guests to new flavours but also educate them about the significance of each dish, enhancing their appreciation of culinary traditions from around the world.

3. Incorporate Symbolic Ingredients and Presentation Techniques

The symbolic meaning behind certain ingredients provides an opportunity for chefs to add depth to their dishes. Ingredients like pomegranates (associated with fertility and abundance), rice (a symbol of prosperity), and honey (representing sweetness and joy) carry special significance in many cultural festivals. Chefs can incorporate these ingredients into their dishes, either as main components or as thoughtful garnishes that add both flavour and meaning.

Presentation techniques can also evoke festive themes. For example, chefs might present dishes in family-style platters to reflect the communal nature of Thanksgiving or Lunar New Year feasts. Using decorative elements, such as edible flowers or symbolic shapes, adds a festive aesthetic to the dish, enhancing the dining experience.

4. Host Themed Pop-Up Events and Seasonal Dinners

Pop-up events and seasonal dinners offer chefs

the chance to create unique, time-limited dining experiences that celebrate specific festivals. For example, a chef might host a Hanami-themed dinner during cherry blossom season, featuring dishes with sakura flavours, or organize a Midsummer celebration with a Scandinavian-inspired menu. These events allow diners to participate in the festivities and experience traditional dishes in a new setting.

Themed events can also include interactive elements, such as live cooking demonstrations or storytelling about the festival's history and significance. This approach enhances the cultural aspect of the meal, providing diners with a deeper appreciation of the festival's culinary traditions.

5. Collaborate with Cultural Experts or Guest Chefs

Collaborating with chefs or experts from different cultural backgrounds can bring authenticity and depth to festive menus. By working alongside individuals who specialize in specific cuisines or cultural festivals, chefs can ensure that their interpretations are respectful and accurate. For instance, a chef might invite a Japanese tea master to co-host a Lunar New Year dinner or work with a Mexican chef to prepare traditional Day of the Dead dishes.

These collaborations not only add authenticity to the menu but also promote cross-cultural exchange, offering diners a genuine taste of different culinary traditions. Guest chefs and cultural experts can share stories and insights, enhancing the overall dining experience with knowledge and respect for the cultural heritage behind each dish.

6. Create Limited-Time Seasonal Specials

Seasonal specials are a great way to introduce festive flavours without overhauling the entire menu. Chefs can offer limited-time dishes that highlight seasonal ingredients or reflect the themes of certain festivals. For example, a restaurant might introduce a spiced cranberry dessert in November for Thanksgiving or a gingerbread-inspired cake during the winter holiday season. These specials allow diners to experience the flavours of the season, creating a sense of anticipation and excitement.

Festivals and seasonal celebrations offer chefs an opportunity to explore the richness of cultural traditions and the symbolism of seasonal ingredients. By incorporating festive elements thoughtfully, chefs can create menus that celebrate global diversity, deepen diners' connection to tradition, and offer a unique and memorable dining experience.

CHAPTER 24: PLANT-BASED EATING AND CULTURAL ATTITUDES TOWARDS VEGETARIANISM

Historical And Cultural Perspectives On Vegetarianism And Plant-Based Diets

Vegetarianism and plant-based diets have deep historical roots and cultural significance across various societies. Although these diets have gained popularity in recent years due to health, environmental, and ethical considerations, plant-based eating has been an integral part of many cultures for centuries. From spiritual beliefs and social norms to environmental factors and personal values, the motivations behind vegetarianism and plant-based diets have varied widely, reflecting the diverse perspectives and practices around the world.

One of the earliest recorded references to vegetarianism is found in ancient India, where the principles of Ahimsa (non-violence) promoted by Hinduism, Buddhism, and Jainism discouraged the consumption of animal flesh. Followers of these religions believed that abstaining from meat was a way to live harmoniously with other creatures, minimizing harm and respecting the sanctity of life. In Hinduism, the cow is particularly revered, and vegetarianism is often seen as a way to honour this sacred relationship. Jainism, one of the most rigorous advocates of non-violence, encourages an entirely plant-based diet, avoiding root vegetables to prevent harm to living organisms in the soil. This spiritual and ethical foundation has made India one of the most predominantly vegetarian societies in the world, with vegetarianism deeply woven into the cultural fabric.

In ancient Greece, philosophers like Pythagoras advocated for a vegetarian lifestyle, viewing it as a means of promoting physical health and mental clarity. Pythagoreanism, as it was known, suggested that abstaining from meat encouraged purity of mind and body, aligning with the philosophical quest for wisdom. Although vegetarianism did not become widespread in ancient Greece, these ideas influenced later generations and contributed to the emergence of vegetarian movements in Western philosophy and literature. Similarly, in ancient Egypt, certain religious groups practiced vegetarianism as part of ritual purity, avoiding animal products as they prepared for sacred rites.

In the modern era, vegetarianism gained prominence during the 19th and early 20th centuries, particularly in Europe and North America, where advocates like Leo Tolstoy, George Bernard Shaw, and Mahatma Gandhi promoted plant-based diets for ethical and health reasons. The vegetarian movement became associated with social reform, environmental awareness, and compassion for animals, inspiring individuals to embrace plant-based diets as a way to live in alignment with values of justice and empathy. This shift laid the foundation for contemporary vegetarian and vegan movements, which continue to advocate for animal welfare, sustainable agriculture, and personal well-being.

Cultural attitudes toward vegetarianism have varied widely, with some societies viewing it as a moral or spiritual duty and others as a personal choice. While certain cultures have embraced vegetarianism as a core aspect of their identity, others have celebrated the abundance of animal products in their cuisine, viewing meat as a symbol of wealth, strength, and vitality. For chefs, understanding these historical and cultural perspectives on plant-based diets provides valuable insight into the motivations and values that shape vegetarian traditions, allowing them to approach plant-based cooking with a deeper appreciation for its cultural significance.

Exploration of Plant-Based Traditions in India, Southeast Asia, and the Mediterranean

The culinary traditions of India, Southeast Asia, and the Mediterranean offer rich examples of plant-based eating, each region bringing unique flavours, techniques, and philosophies to vegetarian cuisine. These traditions demonstrate that plant-based diets can be diverse, flavourful, and satisfying, showcasing the potential of vegetarianism as both a cultural expression and a source of nourishment.

India: A Spiritual and Culinary Heritage of Vegetarianism

India's plant-based culinary traditions are some of the most well-established and diverse in the world. With millions of people practicing vegetarianism, India offers a wealth of vegetarian dishes that highlight its regional flavours, spices, and ingredients. Central to Indian vegetarian cuisine is the use of lentils, chickpeas, vegetables, and dairy products like yoghurt, ghee, and paneer. Staples like dal (lentil stew), sabzi (vegetable stir-fry), and roti (flatbread) are found across the country, with each region incorporating local spices and ingredients to create unique variations.

In the northern regions of India, dishes like palak paneer (spinach with Indian cheese), chole (spiced chickpeas), and aloo gobi (potatoes and cauliflower) showcase the richness of spices such as turmeric, cumin, and coriander. In southern India, vegetarian cuisine features coconut, tamarind, and curry leaves, with dishes like sambar (a spicy lentil stew), rasam (a tangy soup), and idli (steamed rice cakes) that highlight the flavours of the region. The use of spices, herbs, and fermented foods gives Indian vegetarian dishes depth and complexity, demonstrating that plant-based eating can be flavourful and satisfying without meat.

Indian vegetarian cuisine is closely tied to religious practices and the principle of Ahimsa, or non-violence. This cultural emphasis on compassion for all living beings has shaped India's culinary identity, making vegetarianism more than just a dietary choice. For chefs, Indian cuisine offers inspiration on how to create diverse, plant-based dishes that celebrate the versatility of vegetables, legumes, and spices.

Southeast Asia: Plant-Based Flavours and Traditions

Southeast Asia's plant-based traditions reflect the region's abundance of fresh produce, herbs, and spices, as well as the influence of Buddhism, which encourages a plant-based diet. Countries like Thailand, Vietnam, and Indonesia have rich vegetarian culinary traditions that emphasize fresh herbs, tofu, rice, and coconut. Southeast Asian dishes often balance sweet, sour, salty, and spicy flavours, creating vibrant and aromatic meals that are both nourishing and delicious.

In Thailand, dishes like som tam (green papaya salad), tom yum (spicy and sour soup), and pad Thai with tofu highlight the balance of flavours that characterize Thai cuisine. These dishes incorporate ingredients like lime, chili, peanuts, and basil, creating complex flavours that are satisfying without the need for meat. In Vietnam, plant-based dishes such as pho chay (vegetarian noodle soup), goi cuon (fresh spring rolls), and banh mi with tofu reflect the country's emphasis on fresh herbs, vegetables, and rice-based foods.

Indonesia, known for its rich spice heritage, also has a variety of plant-based dishes. Gado-gado, a salad of steamed vegetables with peanut sauce, is a popular vegetarian dish that combines textures and flavours, while tempeh, a fermented soybean product, is widely used as a protein source in dishes like tempeh goreng (fried tempeh). For chefs, Southeast Asia's plant-based cuisine offers a model for creating dishes that are fresh, flavourful, and packed with herbs and spices, showing that plant-based food can be exciting and dynamic.

The Mediterranean: Celebrating the Abundance of the Earth

The Mediterranean region, particularly countries like Greece, Italy, and Lebanon, is renowned for its plant-based dishes that emphasize fresh vegetables, legumes, olive oil, and grains. The Mediterranean diet is celebrated for its health benefits and focus on whole foods, with plant-based dishes playing a central role. Dishes like hummus, tabbouleh, falafel, ratatouille, and pasta e fagioli highlight the region's emphasis on simplicity, freshness, and the quality of ingredients.

In Greece, dishes like spanakopita (spinach pie), fasolada (bean soup), and gigantes plaki (baked giant beans) demonstrate the versatility of legumes and leafy greens, often combined with olive oil, lemon, and herbs like oregano and dill. Italian cuisine also offers numerous vegetarian options, with dishes like pasta primavera (pasta with fresh vegetables), caponata (eggplant salad), and minestrone (vegetable soup) that celebrate seasonal produce.

Lebanon and the broader Middle East also have a rich tradition of plant-based foods, with dishes like mujadara (lentils and rice with caramelized onions), baba ghanoush (eggplant dip), and fattoush (vegetable and bread salad). These dishes are rooted in the region's agricultural heritage, emphasizing the flavours of olives, chickpeas, grains, and fresh herbs. The Mediterranean diet's focus on vegetables, legumes, and whole grains has made it a model for balanced, nutritious eating, showing that plant-based foods can be both simple and deeply satisfying.

For chefs, the plant-based traditions of India, Southeast Asia, and the Mediterranean offer diverse culinary techniques, ingredients, and flavours, demonstrating that plant-based cooking can be as varied and complex as any meat-based cuisine.

Practical Tips for Chefs to Incorporate Plant-Based, Culturally Inspired Dishes

Incorporating plant-based, culturally inspired dishes into a menu offers chefs an opportunity to explore new flavours and celebrate the diversity of global cuisines. By drawing inspiration from traditional plant-based practices, chefs can create dishes that are both delicious and aligned with contemporary dietary preferences. Here are some practical tips for chefs looking to embrace plant-based cooking in a culturally respectful and innovative way.

1. Embrace Legumes, Grains, and Vegetables as the Centrepiece

One of the keys to creating satisfying plant-based dishes is to treat legumes, grains, and vegetables as the main components of the dish, rather than as side items. Chefs can explore the versatility of ingredients like lentils, chickpeas, quinoa, farro, and tofu, creating dishes that highlight their flavours and textures. For example, a chef might create an Indian-inspired dish featuring spiced lentil dal with rice and seasonal vegetables, allowing the lentils to be the focal point.

Incorporating a variety of grains, such as bulgur, millet, and couscous, can also add texture and depth to plant-based meals. Vegetables like eggplant, cauliflower, and mushrooms can be used creatively as the centrepiece of a dish, mimicking the texture and heartiness of meat when prepared with flavourful marinades and spices.

2. Use Cultural Techniques to Enhance Flavour

Many cultures have developed unique techniques for preparing plant-based dishes, such as marinating, fermenting, and spice blending, which can enhance the flavours of vegetables and legumes. Chefs can explore these techniques to create plant-based dishes that are flavourful and satisfying. For instance, using fermentation to make kimchi or tempeh can add depth to a dish, while spice blends like garam masala, za'atar, or berbere bring layers of flavour to vegetables and grains.

Roasting, grilling, and slow-cooking are other techniques that can bring out the natural flavours of vegetables, creating a rich, satisfying taste. By incorporating these methods, chefs can transform simple ingredients into complex, flavourful dishes that are deeply rooted in culinary traditions.

3. Create Plant-Based Versions of Iconic Dishes

Chefs can create plant-based versions of iconic dishes from various cultures, offering diners a familiar yet innovative experience. For example, a chef might serve a vegan "meatball" pasta inspired by Italian cuisine, using chickpeas or mushrooms as a base for the meatballs and pairing it with a rich tomato sauce. Similarly, chefs can make plant-based versions of Middle Eastern falafel wraps, Indian biryani, or Thai green curry, maintaining the flavours of the original dish while using plant-based ingredients.

By reimagining traditional dishes, chefs can introduce diners to plant-based options that are both comforting and exciting, bridging the gap between cultural familiarity and dietary innovation.

4. Incorporate Fresh Herbs and Spices to Add Depth and Aroma

Herbs and spices are essential for creating flavourful plant-based dishes, especially in cuisines where vegetarianism has a strong cultural presence. Chefs can experiment with fresh herbs like basil, cilantro, mint, and dill, as well as spice combinations that add complexity and aroma. For instance, a Mediterranean-inspired dish might use oregano, thyme, and sumac to enhance vegetables, while a Southeast Asian dish could incorporate lemongrass, galangal, and lime leaves for a refreshing taste.

Seasoning vegetables, grains, and legumes with bold flavours adds interest to plant-based dishes, ensuring that they are as satisfying as any meat-based meal. Herbs and spices not only enhance flavour but also provide cultural authenticity, connecting diners with the traditional flavours of each region.

5. Educate Diners on the Cultural Significance of Plant-Based Dishes

Plant-based dishes offer an opportunity for chefs to educate diners about the cultural significance of vegetarian traditions. By providing context and sharing the stories behind each dish, chefs create a more meaningful dining experience. Menus can include brief descriptions about the cultural roots of a dish, such as the Indian tradition of Ahimsa or the Mediterranean emphasis on seasonal ingredients.

Educating diners encourages them to appreciate the history and values behind plant-based dishes, creating a connection that goes beyond taste. This approach fosters a deeper understanding of cultural traditions, highlighting the rich heritage behind each meal.

Incorporating plant-based, culturally inspired dishes into a menu allows chefs to celebrate the diversity of global cuisines while meeting the growing demand for vegetarian and vegan options. By embracing legumes, grains, and vegetables, using traditional techniques, and educating diners, chefs can create plant-based dishes that are flavourful, satisfying, and culturally authentic.

CHAPTER 25: ALCOHOLIC BEVERAGES AND THEIR CULTURAL SIGNIFICANCE

Historical Importance Of Alcoholic Beverages In Different Societies (E.g., Wine In Europe, Sake In Japan)

Alcoholic beverages have held an essential place in human history, with their origins dating back thousands of years. These drinks have been more than mere refreshments; they have served as symbols of cultural identity, ritual offerings, social lubricants, and markers of status. Across different societies, alcoholic beverages have developed unique cultural associations, each drink carrying its own set of customs, meanings, and roles in social and religious life.

Wine has been a significant cultural artifact in Europe

for millennia, particularly in Mediterranean countries like Greece, Italy, and France, where winemaking began as early as 6,000 BCE. In ancient Greece, wine was not only a beverage but also a symbol of civilization and intellectual life. The Greeks held symposia, social gatherings where wine flowed freely, and philosophical discussions unfolded. For the Romans, wine was a symbol of power, pleasure, and hospitality, central to banquets and feasts that reflected the wealth and influence of the host. As Christianity spread through Europe, wine took on religious significance, especially in the context of the Eucharist, where it symbolized the blood of Christ. In France, wine evolved into a national symbol, and the French refined winemaking techniques, establishing themselves as leaders in viticulture. French wine regions like Bordeaux, Burgundy, and Champagne have become synonymous with quality, each region's terroir lending distinct characteristics to its wines.

Sake, Japan's iconic rice wine, has been culturally significant since ancient times. Originating as early as the 3rd century, sake evolved from a simple fermented rice beverage to a drink that embodies Japanese values of craftsmanship, ritual, and respect for nature. During Shinto religious ceremonies, sake is used as an offering to kami (gods), symbolizing purity and reverence. In everyday life, sake is served at celebrations, weddings, and festivals, where it marks the coming together of family and friends. The production of sake reflects Japanese attention to detail and quality, with skilled brewers carefully monitoring fermentation and temperature to create sake with delicate, balanced flavours. Sake breweries are highly respected in Japan, and the drink itself is seen as a connection to Japanese heritage, craftsmanship, and spirituality.

In Central and South America, alcoholic beverages have also played significant roles in cultural and religious practices. Pulque, a fermented drink made from the sap of the agave plant, was central to the social and spiritual life of the Aztecs in ancient Mexico. It was considered a sacred beverage, associated with fertility and the gods, and was often consumed during rituals. Today, pulque remains a traditional drink in Mexico, enjoyed alongside other agave-based spirits like mezcal and tequila, each with its own rich history and cultural importance.

In Europe, Asia, the Americas, and beyond, alcoholic beverages have often served as markers of social identity, distinguishing social classes and providing a way to celebrate, mourn, and connect. For example, vodka in Russia has long been associated with national pride and the communal nature of Russian culture. Drinking vodka is often accompanied by toasts and shared in a communal setting, reflecting the value placed on social bonds. In the Middle East, arak (an anise-flavoured spirit) is commonly enjoyed in Lebanon, Jordan, and Syria, often paired with mezze dishes and consumed during long social gatherings.

The historical significance of alcoholic beverages highlights their deep connections to culture, geography, and tradition. Each drink tells a story of the people who produce and consume it, offering insight into their values, beliefs, and social customs. For chefs, understanding these cultural associations enriches their approach to pairing drinks with food, allowing them to create dining experiences that respect and celebrate the heritage behind each beverage.

Case Studies of Culturally Significant Drinks and Drinking Customs

Alcoholic beverages are often accompanied by unique drinking customs that reflect cultural values and enhance the social experience of sharing a drink. The following case studies illustrate how specific drinks and their associated rituals contribute to cultural identity and social cohesion.

1. Champagne in France

Champagne, a sparkling wine from the Champagne region of France, is more than just a celebratory drink; it symbolizes elegance, prestige, and national pride. Champagne is closely regulated, with only sparkling wines produced in this specific region allowed to bear the name. The drink has become synonymous with celebration, used to mark important life events like weddings, anniversaries, and New Year's Eve. The custom of opening a bottle of champagne with a saber, known as sabrage, dates back to the Napoleonic Wars and adds an element of spectacle to the experience. Champagne's association with joy, luxury, and achievement has made it an integral part of French identity and a beloved choice for toasts around the world.

2. Japanese Sake Etiquette

In Japan, drinking sake is often accompanied by specific customs that reflect respect and social harmony. When sharing sake, it is customary for people to pour sake for each other rather than for themselves, symbolizing a gesture of generosity and humility. In formal settings, the youngest person often pours for their elders, and one should hold the cup with both hands as a sign of respect when receiving a pour. Sake is typically served in small ceramic cups called ochoko and may be warmed or served cold, depending on the season and type of sake. The ritual of pouring and sharing sake fosters a sense of connection and mutual respect, reflecting the importance of courtesy and social cohesion in Japanese culture.

3. Mezcal and Rituals in Mexico

In Mexico, mezcal is a traditional spirit made from the agave plant, often referred to as "the elixir of the gods." Mezcal is deeply rooted in Mexican heritage, and its production is highly artisanal, with many varieties crafted by small, family-owned distilleries. The drinking of mezcal is often accompanied by a ritualistic element, with people following the saying, "Para todo mal, mezcal; para todo bien, también" ("For everything bad, mezcal; for everything good, mezcal too"). Mezcal is typically sipped slowly, allowing the drinker to appreciate its smoky, earthy flavour, and is often served with sal de gusano (worm salt) and a slice of orange. This ritualized drinking process reflects mezcal's cultural significance and its role in social gatherings and celebrations.

4. The Russian Vodka Tradition

In Russia, vodka holds a place of pride and is seen as a drink that brings people together. Vodka is often consumed during long meals, with toasts punctuating each round. These toasts are an essential part of the experience, allowing guests to express wishes for health, happiness, and friendship. It is customary to drink vodka in one gulp, symbolizing strength and resilience, and it is often paired with traditional Russian foods like zakuski (small appetizers). The communal aspect of vodka drinking reflects the importance of camaraderie in Russian culture, where shared drinks are seen as a way to bond and foster trust.

These case studies demonstrate how alcoholic beverages are often intertwined with cultural rituals and social customs, transforming the act of drinking into a meaningful experience. Each drink carries with it a unique set of traditions, enhancing the way people connect with their heritage and each other.

How Chefs Can Pair Foods with Culturally Significant Beverages to Enhance the Dining Experience

For chefs, pairing food with culturally significant beverages provides an opportunity to create immersive and memorable dining experiences. Thoughtful pairings can enhance the flavours of both the food and the drink, celebrating the cultural heritage behind each element. Below are practical tips for chefs on pairing foods with culturally significant beverages to elevate the dining experience.

1. Match Flavours and Complementary Elements

Pairing foods with culturally significant beverages begins with understanding the flavour profiles of both elements. For instance, the acidity and effervescence of champagne make it an excellent pairing for foods with rich, fatty flavours, such as seafood, creamy sauces, or caviar. The crispness of the champagne cuts through the richness, creating a balanced, refreshing experience. Similarly, the smoky and earthy tones of mezcal pair well with robust flavours, such as grilled meats, spicy salsas, or roasted vegetables. By matching complementary flavours, chefs can create harmonious pairings that highlight the unique characteristics of each dish and drink.

2. Embrace Regional Pairings

One of the most effective ways to create culturally authentic pairings is to match foods with beverages from the same region. For example, a Japanese meal featuring sushi, sashimi, or tempura can be beautifully complemented by sake, as the subtle flavours of the sake enhance the delicate taste of the fish and rice. In Italian cuisine, pairing a rich pasta dish with a Tuscan red wine, such as Chianti, brings out the flavours of tomatoes, garlic, and olive oil, creating a cohesive and regionally authentic experience. By embracing regional pairings, chefs honour the culinary traditions of the area and create a sense of place that resonates with diners.

3. Highlight Traditional Drinking Customs

Incorporating traditional drinking customs into the dining experience adds depth and authenticity. For instance, if a chef is serving mezcal, they could introduce the drink with a small ritual, inviting diners to sip slowly and savour the flavours, perhaps even pairing it with a tasting of worm salt and orange slices. For a Russian

-inspired dinner, the chef could introduce a vodka toast between courses, sharing the cultural background and inviting guests to participate in a communal experience. These customs create a connection between the diner and the cultural heritage of the drink, making the meal more engaging and memorable.

4. Use Pairing to Tell a Story

Chefs can use pairings to tell a story about the history and significance of each beverage, educating diners while enhancing the experience. For example, a chef might pair French champagne with a modern take on a traditional French dish, explaining how champagne became synonymous with celebration and luxury in France. This storytelling approach not only enriches the dining experience but also creates an emotional connection to the meal, allowing diners to appreciate the cultural nuances behind each pairing.

5. Experiment with Temperature and Glassware

Serving alcoholic beverages at the appropriate temperature and in the correct glassware enhances their flavour and cultural authenticity. For example, sake can be served warm or cold depending on the season and type, with warm sake pairing well with hearty, comforting dishes in winter, while chilled sake complements lighter, fresher flavours in summer. Glassware also matters; mezcal is traditionally served in small clay cups called copitas, which allow the drinker to fully appreciate the aroma and flavour of the spirit. Paying attention to temperature and glassware enhances the presentation and respects the cultural traditions associated with each beverage.

6. Educate and Engage Diners on the Cultural Context

Educating diners about the cultural background of each beverage creates a more immersive experience. Chefs can share information about the origins, production, and significance of the drink, helping diners understand its role in the culture. This approach turns the meal into a cultural journey, allowing diners to connect with the history and values behind each pairing. For example, a chef might explain the significance of pulque in Mexican culture, enhancing the diners' appreciation of both the drink and the accompanying dish.

Pairing foods with culturally significant beverages offers chefs a way to celebrate heritage, enhance flavours, and create memorable dining experiences. By embracing regional pairings, respecting drinking customs, and educating diners, chefs can honour the cultural richness behind each beverage, transforming the meal into an immersive cultural experience.

CHAPTER 26: FUSION VS. AUTHENTICITY: BALANCING INNOVATION WITH TRADITION

Exploration Of The Fusion Movement And The Debate Over Culinary Authenticity

The fusion movement has emerged as a powerful force in contemporary cuisine, blending ingredients, techniques, and flavours from different cultures to create innovative dishes that push the boundaries of tradition. This approach has been celebrated for its creativity, allowing chefs to combine elements from various culinary backgrounds to create new and exciting flavour profiles. At its best, fusion cuisine represents the evolving nature of food, highlighting how cultural exchange and globalization have enriched the culinary world. However, the fusion movement has also sparked debates about authenticity, cultural appropriation, and the potential dilution of traditional cuisines.

Fusion cuisine has roots in cultural exchanges that span centuries. For instance, the introduction of tomatoes to Italy and chilies to Asia following the Columbian Exchange reshaped these regions' culinary landscapes, leading to the development of beloved dishes like Italian pasta with tomato sauce and spicy Thai curries. However, modern fusion cuisine differs in that it consciously blends cuisines in ways that may not reflect historical or regional connections. This deliberate experimentation has become a hallmark of fusion cuisine, challenging traditional boundaries and inviting diners to explore unexpected pairings.

The fusion movement gained momentum in the late 20th century, with chefs like Wolfgang Puck and Roy Yamaguchi leading the charge in the United States. Puck's blend of French, Californian, and Asian influences in dishes like his famous smoked salmon pizza transformed the American dining scene, showcasing fusion as a form of culinary innovation. This approach resonated with diners looking for new and adventurous flavours, encouraging a wave of chefs to explore cross-cultural pairings that defy tradition.

However, the rise of fusion cuisine has also raised concerns about authenticity and respect for cultural heritage. For some, fusion can be seen as a superficial or commercialized approach that risks diluting or misrepresenting traditional cuisines. Authenticity, in this context, is viewed as the preservation of cultural knowledge, recipes, and techniques that have been passed down through generations. Authentic cuisine is rooted in a specific cultural context, shaped by local ingredients, history, and values that define a community's culinary identity. When traditional dishes are adapted or combined without understanding their cultural significance, there is a risk of oversimplifying or misrepresenting the cuisine.

The debate over fusion and authenticity also intersects with issues of cultural appropriation. Critics argue that certain fusion dishes can be exploitative, borrowing elements from marginalized cultures without acknowledging or respecting their origins. For instance, using ingredients like miso or kimchi without understanding their cultural context can reduce these elements to trendy flavours, disregarding the rich history and traditions behind them. For chefs, this debate presents both a challenge and an opportunity, inviting them to navigate the fine line between innovation and respect.

Balancing fusion with authenticity requires an understanding of the cultural and historical significance of each ingredient and technique. By approaching fusion with intention and respect, chefs can create dishes that honour the heritage behind each element, fostering a deeper appreciation for the diversity of global cuisines.

Case Studies of Successful Fusion Dishes and Approaches (e.g., Peruvian-Japanese Nikkei Cuisine)

Successful fusion cuisine combines elements from different culinary traditions in a way that is thoughtful, respectful, and harmonious. Below are case studies of notable fusion cuisines and dishes that demonstrate how cross-cultural experimentation can create unique and meaningful culinary expressions.

1. Peruvian-Japanese Nikkei Cuisine

One of the most celebrated examples of fusion cuisine is Nikkei, a blend of Peruvian and Japanese culinary traditions. Nikkei cuisine emerged from the Japanese diaspora in Peru, where Japanese immigrants adapted their traditional cooking techniques to the local ingredients available in South America. This fusion developed organically over generations, resulting in dishes that reflect both Peruvian and Japanese influences. Key ingredients in Nikkei cuisine include Peruvian ají peppers, corn, and potatoes, combined with Japanese staples like soy sauce, miso, and sushi-grade fish.

Nikkei dishes such as tiradito, a raw fish dish similar to ceviche but sliced thinly like sashimi, embody this fusion. Tiradito is typically served with Peruvian spices and flavours, like lime and ají amarillo, creating a dish that bridges both culinary traditions. Another popular Nikkei dish is causa sushi, which incorporates the Peruvian potato dish causa with sushi-style presentation. Nikkei cuisine illustrates how fusion can occur naturally, blending flavours and techniques in a way that respects the cultural heritage of both culinary traditions.

2. Korean-Mexican Fusion: Korean Tacos

Korean-Mexican fusion, popularized by chefs like Roy Choi, is another successful example of fusion cuisine. Choi's Kogi BBQ Tacos, which combine Korean marinated meats with Mexican tortillas and toppings, have become an iconic street food in Los Angeles. The combination of spicy Korean barbecue flavours with the portable, handheld appeal of Mexican tacos has resonated with diners, showcasing how two vibrant street food cultures can come together in a single dish.

Korean tacos incorporate ingredients like bulgogi (marinated beef), kimchi, and gochujang (Korean chili paste) within a tortilla, blending the rich flavours of Korean cuisine with the Mexican tradition of taco-making. This fusion approach respects both culinary traditions, celebrating their bold flavours and communal nature. Korean-Mexican fusion has inspired further experimentation, with chefs creating dishes like kimchi quesadillas and bibimbap burritos, each highlighting the adaptability of Korean and Mexican ingredients.

3. Indian-Chinese Fusion (Chindian Cuisine)

Indian-Chinese fusion, often referred to as Chindian cuisine, is popular in India, where it combines Chinese cooking techniques with Indian spices and flavours. Chindian cuisine developed as Chinese immigrants adapted their cuisine to Indian tastes, resulting in dishes that are neither fully Indian nor fully Chinese. Dishes like chili chicken, Manchurian, and Hakka noodles are staples in Indian-Chinese restaurants, blending the spiciness of Indian food with the stir-frying techniques of Chinese cuisine.

Manchurian, a popular Chindian dish, consists of deep-fried vegetable or meat balls in a spicy, tangy sauce that combines soy sauce, ginger, garlic, and Indian spices. The dish's bold flavours and rich texture have made it a favourite in India, where it reflects both the adaptability and creativity of fusion cuisine. Chindian cuisine is a testament to how fusion can evolve through cultural interaction, resulting in new and exciting flavours that reflect the culinary heritage of both traditions.

These case studies illustrate that successful fusion is possible when chefs understand and respect the origins of each cuisine. By blending techniques and ingredients thoughtfully, fusion dishes can become a celebration of cultural diversity and culinary innovation.

Tips for Chefs on Navigating Fusion While Respecting Cultural Origins

For chefs, navigating the fusion movement while respecting cultural origins requires a mindful approach that balances creativity with cultural understanding. Here are practical tips for chefs who want to experiment with fusion cuisine responsibly and authentically.

1. Research the Cultural and Historical Background of Each Cuisine

Before blending elements from different culinary traditions, it is essential for chefs to research the cultural and historical background of each cuisine. Understanding the significance of ingredients, techniques, and presentation methods helps chefs approach fusion with respect and depth. For example, if a chef wants to create a Japanese-Mexican fusion dish, they could learn about the importance of umami flavours in Japanese cuisine and the Mexican emphasis on bold spices, allowing them to combine these elements thoughtfully. This research fosters a deeper appreciation for the cultural heritage behind each cuisine and helps chefs avoid superficial or inappropriate combinations.

2. Highlight the Authenticity of Each Component

When creating fusion dishes, chefs can celebrate the authenticity of each component by allowing its unique flavours and characteristics to shine. Rather than blending flavours indiscriminately, chefs can create dishes that maintain the integrity of each ingredient or technique. For example, in Nikkei cuisine, the Japanese influence is evident in the preparation of raw fish, while Peruvian ingredients like ají peppers add local flavour. By emphasizing the authenticity of each component, chefs create fusion dishes that feel balanced and harmonious.

3. Use Fusion as a Medium for Storytelling

Fusion cuisine can be a powerful way to tell a story about cultural exchange, migration, or shared culinary heritage. Chefs can use fusion to reflect their personal experiences, cultural identity, or travels, creating dishes that are meaningful and intentional. For example, a chef with a multicultural background might create a fusion dish that represents their heritage, blending flavours that are personally significant. This approach adds depth to fusion cuisine, transforming it from a trend into a medium for storytelling and cultural expression.

4. Avoid Tokenism and Respect Cultural Sensitivity

Tokenism, or using culturally significant ingredients and techniques without understanding their context, can reduce fusion to a superficial exercise. To avoid tokenism, chefs should be mindful of the cultural sensitivity surrounding certain ingredients or cooking methods. For example, using sacred or culturally significant ingredients, such as turmeric in Indian cuisine or miso in Japanese cuisine, requires respect and understanding. When using these ingredients, chefs can acknowledge their cultural significance, ensuring that fusion dishes are created with appreciation rather than appropriation.

5. Collaborate with Experts or Cultural Ambassadors

Collaboration can enhance authenticity in fusion cuisine. Chefs can work with experts or cultural ambassadors from the cuisines they wish to blend, gaining insights into traditional methods, flavour profiles, and cultural nuances. For example, a chef interested in blending Peruvian and Japanese flavours might collaborate with a Nikkei chef to learn about the balance of flavours in Nikkei cuisine. Collaboration fosters cross-cultural exchange and enriches the fusion process, ensuring that dishes are crafted with genuine respect and authenticity.

6. Balance Creativity with Cultural Integrity

Creativity is essential in fusion cuisine, but it should be balanced with cultural integrity. When experimenting with flavours, chefs can focus on complementary elements rather than forcing unfamiliar combinations. For example, combining Italian and Thai flavours might seem unconventional, but finding common ground—such as the use of basil in both cuisines—can create a harmonious fusion. This approach respects the essence of each cuisine while allowing room for innovation, creating dishes that are both inventive and grounded.

7. Educate Diners About the Cultural Context of Fusion Dishes

Educating diners about the cultural background of fusion dishes enhances the dining experience and fosters a greater appreciation for the culinary heritage behind each element. Chefs can share stories about the origins of ingredients, the inspiration for the dish, and the significance of each cultural component. This approach transforms fusion cuisine into an opportunity for cultural education, allowing diners to engage with the dish on a deeper level.

Fusion cuisine can be a celebration of cultural diversity and culinary creativity when approached thoughtfully. By researching cultural backgrounds, highlighting authenticity, avoiding tokenism, and collaborating with experts, chefs can create fusion dishes that honour tradition while embracing innovation. This balance between fusion and authenticity enriches the culinary world, offering diners unique flavours that respect and celebrate the heritage behind each cuisine.

CHAPTER 27: FOOD AND THE ENVIRONMENT

Overview Of The Relationship Between Food Practices And Environmental Impact

The relationship between food and the environment is one of profound interdependence, where the methods of growing, producing, processing, and consuming food have significant implications for the health of the planet. Agriculture, which supports the world's food supply, is a leading contributor to environmental issues, including deforestation, water pollution, biodiversity loss, and greenhouse gas emissions. As the global population grows and demand for food rises, the impact of food production on the environment has become a critical area of concern, urging societies to adopt more sustainable and eco-friendly practices.

One of the most pressing environmental challenges associated with food production is deforestation. Forests are often cleared to make way for agricultural land, particularly for crops like soy, palm oil, and cattle ranching. This deforestation contributes to habitat loss, endangering biodiversity, and disrupts carbon storage, as trees play a vital role in absorbing and storing carbon dioxide. The loss of forests, particularly in tropical regions like the Amazon, accelerates climate change and threatens the delicate balance of ecosystems that support a diverse range of plant and animal species.

Another environmental issue closely linked to food production is water use and pollution. Agriculture is responsible for approximately 70% of global freshwater withdrawals, with certain crops, such as rice and almonds, requiring significant amounts of water to grow. Irrigation, while essential for farming, can lead to water scarcity in regions where resources are already limited, affecting local communities and ecosystems. Additionally, the use of synthetic fertilizers and pesticides often results in runoff that pollutes rivers, lakes, and oceans, harming aquatic life and disrupting natural water systems.

Greenhouse gas emissions from agriculture, including methane and nitrous oxide, contribute to global warming. Livestock farming, particularly cattle, produces methane through enteric fermentation, while the use of nitrogen-based fertilizers releases nitrous oxide, a potent greenhouse gas. Together, livestock farming and crop production account for nearly a quarter of global greenhouse gas emissions, making agriculture a significant contributor to climate change.

Food waste is another critical issue with environmental implications. Approximately one-third of all food produced globally is lost or wasted, which translates to wasted resources, energy, and emissions associated with production and disposal. When food waste ends up in landfills, it decomposes and releases methane, further contributing to greenhouse gas emissions.

The environmental impact of food practices highlights the urgent need for sustainable agriculture, responsible sourcing, and mindful consumption. For chefs, understanding this relationship is essential, as the choices they make regarding ingredients, portion sizes, and waste management can significantly influence their kitchen's environmental footprint. By embracing sustainable practices, chefs play a vital role in promoting food systems that respect and protect the environment.

Sustainable Practices Across Different Cultures (e.g., Indigenous Farming Techniques, Permaculture)

Many cultures have long-standing traditions of sustainable agriculture and resource management, offering valuable insights into eco-friendly practices that promote environmental stewardship. From indigenous farming techniques to modern approaches like permaculture, these practices demonstrate that sustainable food systems are not new concepts but have deep roots in cultural heritage.

Indigenous Farming Techniques

Indigenous communities worldwide have developed agricultural methods that are deeply connected to the land, focusing on sustainability, biodiversity, and respect for natural cycles. These techniques, often passed down through generations, reflect a profound understanding of ecosystems and an emphasis on harmony between people and nature. For instance, the "Three Sisters" method used by Native American tribes like the Iroquois involves planting corn, beans, and squash together. This intercropping technique is sustainable because each plant supports the others: corn provides a structure for beans to climb, beans fix nitrogen in the soil, and squash covers the ground to retain moisture and prevent weeds. This method enhances soil health, reduces the need for artificial fertilizers, and promotes a balanced ecosystem.

In the Andes, indigenous farmers have cultivated crops like potatoes and quinoa for centuries using terrace farming. By building terraces along mountain slopes, they reduce soil erosion, conserve water, and create microclimates that support diverse crops. This practice not only preserves the integrity of the land but also helps adapt to challenging mountainous environments. Similarly, the chinampas system used by the Aztecs in ancient Mexico involves creating floating gardens on shallow lake beds, providing a self-sustaining method of cultivation that conserves water and supports biodiversity.

Indigenous farming techniques emphasize biodiversity, soil health, and minimal waste, showing that sustainable practices are rooted in respect for natural resources. These approaches challenge modern agricultural practices that prioritize high yields and monocultures, offering alternative methods that enhance ecological resilience.

Permaculture

Permaculture is a modern approach to agriculture that draws inspiration from natural ecosystems, aiming to create self-sustaining, regenerative food systems. Developed in the 1970s by Australians Bill Mollison and David Holmgren, permaculture combines the words "permanent" and "agriculture," reflecting the goal of establishing sustainable practices that can endure over time. Permaculture principles prioritize diversity, soil health, water conservation, and minimal intervention, often employing practices like crop rotation, composting, and natural pest control.

Permaculture farms are designed to mimic natural landscapes, with different plants and animals contributing to the system's overall health. For example, animals like chickens or ducks are integrated into the farm to manage pests, fertilize the soil, and provide eggs. Companion planting is also a key aspect of permaculture, where specific plants are grown together to support each other's growth and reduce the need for synthetic fertilizers or pesticides. The focus on closed-loop systems minimizes waste and enhances sustainability, creating food systems that work in harmony with nature rather than against it.

Permaculture has gained popularity globally, with small farms, urban gardens, and community projects adopting its principles. By promoting a regenerative approach to agriculture, permaculture offers an alternative to industrial farming, encouraging practices that enrich the soil, conserve water, and reduce carbon emissions. For chefs interested in sustainable sourcing, permaculture farms provide an eco-friendly option, offering seasonal, organically grown produce that aligns with environmental values.

These sustainable practices from different cultures demonstrate that eco-friendly food production is possible when respect for nature is prioritized. By incorporating principles from indigenous farming, permaculture, and other sustainable methods, chefs can make informed choices that support environmental stewardship and honour cultural traditions of land care.

How Chefs Can Implement Eco-Friendly Practices and Source Ingredients Responsibly

For chefs, adopting eco-friendly practices and sourcing ingredients responsibly goes beyond trend; it reflects a commitment to sustainable food systems and environmental responsibility. By making mindful choices in the kitchen, chefs can reduce waste, conserve resources, and promote ethical sourcing. Here are some practical strategies for chefs looking to minimize their environmental impact.

1. Source Local and Seasonal Ingredients

Using local and seasonal ingredients is one of the most effective ways for chefs to reduce their kitchen's environmental footprint. Locally sourced ingredients require less transportation, reducing greenhouse gas emissions associated with long-distance shipping. Seasonal produce is also fresher and more flavourful, offering diners the best of what nature provides during each season. By working with local farmers and producers, chefs support the regional economy and encourage sustainable agricultural practices within their community. Seasonal menus not only highlight the beauty of fresh produce but also connect diners with the rhythms of nature, emphasizing the importance of eating in harmony with the seasons.

2. Embrace Nose-to-Tail and Root-to-Stem Cooking

Nose-to-tail and root-to-stem cooking are sustainable approaches that emphasize using every part of the ingredient, reducing waste, and respecting the resources involved in food production. Nose-to-tail cooking involves utilizing all parts of the animal, from prime cuts to offal, while root-to-stem cooking applies the same principle to vegetables, using stems, leaves, and peels that are often discarded. Chefs can incorporate these approaches by creating dishes that feature lesser-used cuts of meat, vegetable tops, and skins, finding creative ways to turn scraps into flavourful components. This approach not only minimizes waste but also offers diners a unique culinary experience that celebrates the ingredient in its entirety.

3. Reduce Food Waste Through Smart Menu Planning

Effective menu planning can help reduce food waste by ensuring that ingredients are used efficiently and creatively. Chefs can design menus that repurpose ingredients across different dishes, using leftover trimmings for stocks, sauces, or garnishes. For example, vegetable trimmings can be used to make broth, while excess herbs can be turned into pesto or infused oils. Chefs can also monitor portion sizes, reducing waste by serving reasonable amounts and encouraging diners to enjoy every part of the dish. Some restaurants have implemented "zero-waste" menus, where each dish is designed to use ingredients fully, demonstrating a commitment to sustainability and resourcefulness.

4. Opt for Eco-Friendly Packaging and Minimize Single-Use Items

Incorporating eco-friendly packaging and minimizing single-use items is essential for reducing plastic waste. Many chefs are choosing compostable or biodegradable packaging materials, such as cardboard or plant-based plastics, to reduce their environmental impact. Reusable containers for takeout and delivery orders, as well as wooden or bamboo cutlery, are also sustainable alternatives. In the kitchen, chefs can reduce plastic use by storing ingredients in glass containers or reusable bags, minimizing waste while creating a more sustainable workflow. By opting for environmentally friendly packaging, chefs demonstrate their commitment to reducing plastic pollution and protecting the planet.

5. Source Sustainably Farmed and Ethically Produced Ingredients

Responsible sourcing includes selecting ingredients that are sustainably farmed and ethically produced, with attention to the environmental and social impact of each product. For seafood, chefs can use guides from organizations like the Marine Stewardship Council (MSC) to choose sustainably caught fish, avoiding species that are overfished or farmed unsustainably. For meats, chefs can work with suppliers who prioritize animal welfare, offering grass-fed, pasture-raised, or organic options. Certified fair trade products, such as coffee, chocolate, and tea, ensure that farmers are paid fair wages and work in safe conditions, supporting ethical labour practices. By prioritizing sustainable sourcing, chefs contribute to a food system that respects the planet and its people.

6. Educate Diners and Staff on Sustainable Practices

Educating diners and staff on sustainable practices fosters a culture of environmental awareness and responsibility. Chefs can use menus, signage, or conversations with diners to share the story behind each ingredient, explaining why certain choices—such as using local produce or offering smaller portions —align with sustainability. In the kitchen, chefs can train staff on waste reduction, energy conservation, and responsible sourcing, creating a team that values eco-friendly practices. By encouraging both diners and staff to embrace sustainability, chefs help build a community that supports environmental stewardship.

7. Partner with Food Recovery Programs and Composting Initiatives

Food recovery programs allow restaurants to donate excess food to organizations that feed those in need, reducing food waste while supporting local communities. Chefs can partner with food recovery organizations to ensure that surplus food is redirected rather than discarded. Composting is another way to manage waste sustainably, turning food scraps into nutrient-rich soil. Many restaurants have implemented composting programs, either in-house or through local composting services, reducing the amount of waste sent to landfills. By engaging in food recovery and composting, chefs contribute to a circular economy that minimizes waste and supports community resilience.

Chefs have a unique role in promoting sustainable food practices by implementing eco-friendly methods in the kitchen and sourcing ingredients responsibly. From supporting local farmers to reducing waste and educating diners, each choice chefs make reflects a commitment to a more sustainable food system. These practices not only reduce the environmental impact of the kitchen but also inspire a new generation of diners to appreciate and support eco-conscious dining.

CHAPTER 28:
THE FUTURE OF GASTRONOMY: TRENDS AND INNOVATIONS

Examination Of Emerging Trends In Gastronomy, Such As Edible Insects, Lab-Grown Meat, And Zero-Waste Cooking

The culinary landscape is evolving rapidly, driven by technological advancements, changing consumer preferences, and a growing awareness of sustainability. As the industry shifts toward more eco-conscious and health-oriented approaches, chefs are embracing innovative techniques and ingredients that challenge traditional notions of food. Emerging trends like edible insects, lab-grown meat, and zero-waste cooking exemplify the forward-thinking approach of modern gastronomy, where creativity meets responsibility.

Edible Insects

In recent years, edible insects have gained popularity as a sustainable protein source. Insects like crickets, mealworms, and grasshoppers are rich in protein, vitamins, and minerals, making them a nutritious alternative to conventional meats. They have a small environmental footprint, requiring significantly less water, land, and feed compared to livestock. In many cultures, particularly across Asia, Africa, and Latin America, insects have been consumed for centuries as a traditional source of nutrition. However, in Western cultures, the idea of eating insects is still met with hesitation, as it challenges cultural perceptions of food.

As global demand for protein rises and concerns about the environmental impact of meat production increase, edible insects offer a promising solution. Companies specializing in insect-based foods are experimenting with products like cricket flour, protein bars, and even insect-based snacks, introducing consumers to insect protein in palatable forms. For chefs, incorporating edible insects into their menus presents an opportunity to offer sustainable and nutritious dishes that align with eco-conscious dining trends. Chefs can creatively prepare insects by using them in sauces, salads, and even desserts, introducing diners to this protein source in ways that are accessible and appealing.

Lab-Grown Meat

Lab-grown meat, also known as cultured or cell-based meat, represents a cutting-edge innovation in the food industry. This technology involves cultivating animal cells in a lab to create meat without the need for raising and slaughtering animals. Lab-grown meat addresses many of the environmental and ethical concerns associated with conventional meat production, as it eliminates animal suffering, reduces greenhouse gas emissions, and requires fewer resources. Companies like Memphis Meats and Mosa Meat have pioneered the development of lab-grown meat, and as the technology advances, it is expected to become more affordable and widely available.

The introduction of lab-grown meat has the potential to transform the food industry, offering a more sustainable alternative to traditional meat. While cultured meat is still in its early stages, it has already sparked interest from chefs and food enthusiasts who see its potential to revolutionize the way people consume protein. For chefs, working with lab-grown meat offers a unique opportunity to create innovative dishes that appeal to environmentally conscious diners. As this technology becomes more accessible, it could pave the way for a future where lab-grown meat coexists with plant-based and conventional options, providing greater choice and sustainability.

Zero-Waste Cooking

Zero-waste cooking has become a prominent trend in gastronomy, driven by the need to reduce food waste and promote sustainability. This approach encourages chefs to use every part of an ingredient, minimizing waste and finding creative ways to repurpose scraps and leftovers. Zero-waste cooking aligns with the circular economy model, which aims to reduce waste, reuse resources, and create closed-loop systems. For example, vegetable trimmings can be used to make stocks, while fruit peels can be transformed into syrups or garnishes. By embracing zero-waste principles, chefs contribute to a more sustainable food system and inspire diners to consider the environmental impact of their meals.

Chefs around the world are adopting zero-waste practices in their kitchens, creating menus that prioritize resourcefulness and reduce food waste. Some restaurants have even developed zero-waste menus, where each dish is crafted to use ingredients fully. This approach not only reduces the environmental footprint of the kitchen but also challenges chefs to think creatively, turning byproducts into flavourful and innovative components. Zero-waste cooking offers chefs an opportunity to lead by example, promoting sustainable dining practices that resonate with environmentally conscious diners.

These trends—edible insects, lab-grown meat, and zero-waste cooking—represent the future of gastronomy, where innovation is guided by a commitment to sustainability and ethical responsibility. For chefs, embracing these trends offers a way to stay at the forefront of culinary innovation while addressing some of the most pressing challenges facing the food industry.

Cultural Implications of These Trends and Their Potential Future Impact

The cultural implications of emerging trends in gastronomy are significant, as they challenge long-standing beliefs, culinary traditions, and social norms surrounding food. Each trend introduces new ideas about what food can be, how it can be produced, and what role it plays in society. As these innovations gain traction, they have the potential to reshape the culinary landscape, influencing everything from consumer habits to agricultural practices.

Edible Insects and Cultural Acceptance

The incorporation of edible insects into Western diets challenges traditional ideas about what is considered food. Insects have long been consumed in many cultures as a reliable protein source, but in Western countries, they are often perceived as unappetizing or even repulsive. Overcoming this cultural barrier requires a shift in mindset, as well as creative approaches from chefs to make insects more palatable and familiar. As insect-based products become more common, consumers may gradually become more open to incorporating insects into their diets, recognizing their nutritional and environmental benefits.

In some ways, the trend of edible insects highlights the cultural relativity of food norms. What is considered acceptable in one culture may be viewed differently in another. As globalization brings diverse culinary practices into contact, edible insects may serve as a symbol of cross-cultural exchange and the evolution of food norms. If embraced, insects could become a mainstream protein source, reducing the environmental impact of conventional livestock and creating a more sustainable global food system.

Lab-Grown Meat and Ethical Considerations

Lab-grown meat represents a fundamental shift in how meat is produced, challenging traditional methods and raising ethical questions. For many, the concept of cultured meat aligns with ethical values by providing an option that does not involve animal slaughter. However, it also raises questions about the "naturalness" of food, as lab-grown meat is produced in a controlled environment rather than through traditional farming. For some, this departure from natural production methods may be unsettling, leading to resistance or scepticism.

The introduction of lab-grown meat also has implications for traditional farming communities, as it could disrupt the demand for conventional livestock. This shift may require a reimagining of the agricultural sector, with farmers potentially adopting new roles or technologies to stay relevant. If lab-grown meat becomes widely accepted, it could contribute to a more sustainable food system by reducing the environmental impact of meat production, but it will likely require ongoing dialogue about ethics, transparency, and regulation to address consumer concerns.

Zero-Waste Cooking and the Redefinition of Culinary Values

Zero-waste cooking is redefining the values associated with cooking, emphasizing resourcefulness, creativity, and environmental responsibility. This trend challenges the notion that food byproducts are merely waste, encouraging chefs and consumers to view every part of an ingredient as valuable. By adopting zero-waste principles, chefs not only reduce their kitchen's environmental impact but also promote a culture of mindfulness and appreciation for resources.

The cultural impact of zero-waste cooking extends to diners, who are increasingly aware of the importance of sustainability in food choices. This trend may lead to a shift in consumer expectations, with more diners seeking restaurants that prioritize eco-friendly practices. Zero-waste cooking aligns with broader movements toward sustainable living and responsible consumption, positioning the culinary industry as a leader in promoting environmental consciousness. As this approach becomes more mainstream, it could transform the way food is valued and consumed, fostering a culture that respects resources and minimizes waste.

Each of these trends has the potential to influence future food practices, encouraging greater sustainability, ethical consideration, and cultural exchange. As chefs adopt these trends, they play a vital role in shaping a future where food is produced, prepared, and consumed in ways that honour both people and the planet.

Practical Considerations for Chefs Looking to Adopt Innovative Techniques Responsibly

For chefs, adopting new trends and techniques requires a thoughtful approach that balances innovation with responsibility. By considering the ethical, environmental, and cultural implications of their choices, chefs can incorporate emerging trends in ways that enhance the dining experience while promoting sustainable and respectful practices. Here are some practical considerations for chefs looking to embrace these innovations responsibly.

1. Introduce New Ingredients Gradually and Transparently

When incorporating unconventional ingredients like edible insects or lab-grown meat, transparency is key to building trust with diners. Chefs can introduce these ingredients gradually, starting with small additions to familiar dishes. For example, cricket flour can be used in baked goods, or insect-based protein powders can be blended into sauces or soups. Educating diners about the nutritional benefits and environmental advantages of these ingredients helps them understand the rationale behind the choice.

Chefs can also offer tasting events or sample menus to allow diners to try these ingredients in a comfortable setting. Transparency and openness create a positive experience, encouraging diners to explore new flavours while understanding the motivation behind these choices.

2. Balance Innovation with Culinary Traditions

While embracing trends like lab-grown meat or fusion cuisine, it is essential for chefs to balance innovation with respect for culinary traditions. This balance involves blending new ingredients or techniques in ways that complement, rather than overshadow, the cultural heritage of the cuisine. For example, a chef might create a lab-grown meat dish that incorporates traditional spices or cooking methods from a specific culture, preserving the dish's authenticity while introducing new elements.

Respecting culinary traditions adds depth to innovative dishes, allowing chefs to create meals that honour both heritage and progress. This approach fosters a sense of continuity, demonstrating that innovation can coexist with cultural appreciation.

3. Invest in Training and Education for Zero-Waste Techniques

Zero-waste cooking requires training, as it involves using parts of ingredients that are often discarded. Chefs can invest in training for their staff to learn techniques like making stocks from vegetable trimmings, fermenting fruit peels, or creating powders from dried herb stems. Educating the kitchen team on the principles of zero-waste cooking fosters a culture of creativity and resourcefulness, allowing chefs to maximize the value of every ingredient.

By empowering the kitchen team with zero-waste skills, chefs build a more sustainable operation and encourage a mindset that values each component. Zero-waste cooking not only reduces costs but also enhances the menu with unique flavours created from parts that would otherwise be wasted.

4. Foster Collaboration with Producers and Suppliers

Building relationships with producers and suppliers who share a commitment to sustainability is essential for adopting innovative practices. For example, sourcing edible insects or lab-grown meat requires finding suppliers that prioritize ethical production methods and high-quality standards. Collaborating with sustainable suppliers ensures that chefs have access to responsibly sourced ingredients that align with their values.

Working closely with producers also provides chefs with insight into the sourcing process, allowing them to make informed choices and educate their diners about the origins of each ingredient. This transparency builds trust with diners and strengthens the chef's commitment to responsible sourcing.

5. Educate Diners on the Environmental and Ethical Benefits of Innovation

Educating diners about the environmental and ethical benefits of new trends enhances their dining experience and fosters a sense of connection with the meal. Menus, waitstaff, and digital platforms can all be used to provide information on the impact of choices like edible insects or lab-grown meat. By sharing the sustainability advantages of zero-waste practices, chefs encourage diners to appreciate the efforts made to reduce waste and promote eco-friendly dining.

Adopting innovative techniques responsibly involves a balance of education, transparency, and respect for tradition. By introducing new ingredients gradually, fostering collaboration, and educating diners, chefs can create an immersive dining experience that celebrates the future of gastronomy while honouring cultural and environmental values.

CHAPTER 29: FOOD AS MEDICINE: CULINARY PRACTICES FOR HEALTH AND WELLNESS

Historical Use Of Food As Medicine Across Various Cultures (E.g., Ayurveda, Traditional Chinese Medicine)

The concept of "food as medicine" has been integral to numerous cultural and healing practices throughout history. Many ancient civilizations recognized the healing power of food, developing dietary practices and philosophies that emphasize the connection between diet and well-being. From Ayurvedic principles in India to Traditional Chinese Medicine (TCM) and Indigenous knowledge systems, these traditions view food as a powerful tool to support health, prevent illness, and restore balance in the body.

Ayurveda

Ayurveda, a 5,000-year-old system of medicine from India, places great importance on diet as a means of achieving and maintaining balance within the body. Rooted in the concept of the doshas—Vata, Pitta, and Kapha—Ayurveda views each individual as having a unique constitution that requires specific dietary practices to support health. According to Ayurvedic principles, food not only nourishes the body but also affects the mind, emotions, and spirit. Each dosha is associated with particular elements: Vata with air and ether, Pitta with fire and water, and Kapha with earth and water. Foods are classified by their effects on the doshas, and balancing these elements is believed to be essential for optimal health.

Ayurvedic dietary practices emphasize the use of spices like turmeric, ginger, cumin, and fennel, which are believed to aid digestion, support immunity, and promote balance. For instance, turmeric is prized for its anti-inflammatory properties, while ginger is known for its ability to aid digestion and stimulate circulation. The Ayurvedic approach encourages mindful eating, seasonal adjustments to the diet, and the use of whole, minimally processed foods that align with one's dosha. This philosophy embodies the holistic view that food is a form of preventive medicine, capable of promoting health and longevity.

Traditional Chinese Medicine (TCM)

Traditional Chinese Medicine (TCM) also views food as a fundamental aspect of health, emphasizing the concept of Qi (vital energy) and the balance of Yin and Yang. In TCM, foods are classified by their energy properties (warming, cooling, neutral) and flavours (sweet, sour, bitter, salty, pungent), each of which is believed to have specific effects on the body. Warming foods, such as ginger and garlic, are thought to invigorate the Qi and are beneficial for people with cold symptoms, while cooling foods like cucumber and watermelon help reduce internal heat and soothe inflammation.

TCM dietary practices also align with the seasons, promoting ingredients that support the body's natural cycles. For instance, in winter, warming foods and hearty soups are recommended to nourish and protect the body, while summer calls for cooling, hydrating foods like leafy greens and melons. The use of herbs, roots, and spices in TCM underscores the belief that food can restore balance, strengthen immunity, and prevent illness. Many popular dishes in China, such as chicken soup with ginseng or red date tea, are based on TCM principles, combining ingredients that promote health and harmony.

Indigenous Knowledge Systems

Indigenous cultures around the world also view food as medicine, with dietary practices rooted in a deep connection to the land and natural cycles. For example, Native American tribes traditionally used foods like wild berries, corn, beans, and squash—known as the "Three Sisters"—for their nutritional and healing properties. Indigenous knowledge systems emphasize the importance of seasonal eating, foraging, and sustainable practices, believing that the Earth provides what is needed for each season.

The use of plants for medicinal purposes is common in Indigenous cultures, where foods like cactus, seaweed, and wild herbs are used not only for nourishment but also for healing specific ailments. For instance, some Native American communities use blue cornmeal for its nutrient-dense properties, while seaweed is revered by coastal Indigenous groups for its iodine content and mineral-rich profile. Indigenous food practices reflect a holistic approach to health, where food is valued for its capacity to sustain life, promote wellness, and heal the body.

These historical perspectives on food as medicine highlight the shared belief across cultures that food has the power to support health and prevent disease. For chefs, understanding these traditions provides insight into the healing potential of ingredients and encourages a more mindful, health-oriented approach to cooking.

Case Studies of Health-Focused Cuisines and Their Philosophies (e.g., Mediterranean Diet, Japanese Washoku)

Certain regional cuisines are celebrated for their health-promoting properties, offering balanced and nutritious approaches to eating. The Mediterranean diet and Japanese washoku exemplify health-focused culinary philosophies that emphasize whole foods, balance, and mindful eating.

The Mediterranean Diet

The Mediterranean diet, inspired by the traditional eating habits of countries bordering the Mediterranean Sea, is widely regarded as one of the healthiest diets in the world. This dietary approach emphasizes fresh vegetables, fruits, whole grains, legumes, nuts, seeds, olive oil, and moderate amounts of fish and poultry. Red meat and processed foods are limited, and meals are often enjoyed with family, encouraging mindful, social eating.

One of the core principles of the Mediterranean diet is its emphasis on heart-healthy fats from olive oil and nuts, which are rich in monounsaturated fats and antioxidants. Studies have shown that the Mediterranean diet can lower the risk of cardiovascular disease, improve cognitive health, and support longevity. Key components like tomatoes, garlic, onions, and fresh herbs provide vitamins, minerals, and anti-inflammatory compounds, while red wine, consumed in moderation, contributes antioxidants that may benefit heart health.

The Mediterranean diet's focus on whole foods and simplicity resonates with modern health philosophies, where minimally processed ingredients and seasonal produce form the foundation of nutritious meals. For chefs, incorporating Mediterranean principles into their cooking allows them to create flavourful, balanced dishes that align with health-conscious dining trends.

Japanese Washoku

Washoku, the traditional dietary culture of Japan,

emphasizes harmony and balance, reflecting the Japanese philosophy of Ichiju Sansai, or "one soup, three sides." This approach involves serving a variety of dishes in small portions, each contributing to a balanced meal. Washoku emphasizes seasonal ingredients, rice, vegetables, tofu, seafood, and fermented foods like miso and pickles, which are rich in probiotics that support gut health.

In Japanese cuisine, there is a strong focus on umami, the fifth taste, which enhances flavour and reduces the need for excess salt and sugar. Ingredients like seaweed, shiitake mushrooms, and soy sauce are natural sources of umami and are commonly used to season dishes. Traditional Japanese meals also incorporate a variety of textures, colours, and flavours, contributing to a visually pleasing and nutritionally balanced plate.

The Japanese diet is associated with longevity and low rates of obesity, partly due to its emphasis on fresh, low-calorie foods and portion control. Chefs can draw inspiration from washoku by creating meals that focus on balance, incorporating fermented ingredients, seasonal produce, and umami-rich flavours to offer diners a satisfying and healthful experience.

These health-focused cuisines demonstrate how dietary practices rooted in balance, diversity, and whole foods can promote wellness. For chefs, understanding the philosophies behind these diets offers a foundation for creating nutritious dishes that support health-conscious diners.

Tips for Chefs on Creating Nutritious, Culturally

Inspired Dishes for Health-Conscious Diners

For chefs, incorporating health-focused principles into their cooking allows them to create dishes that not only satisfy the palate but also support well-being. Here are practical tips for chefs on crafting nutritious, culturally inspired dishes that cater to health-conscious diners.

1. Prioritize Whole, Minimally Processed Ingredients

Using whole, minimally processed ingredients is key to creating nutritious dishes. Fresh vegetables, fruits, whole grains, and legumes provide essential nutrients, fibre, and antioxidants, supporting a balanced and healthful diet. Chefs can emphasize whole foods by designing menus that celebrate the natural flavours and textures of seasonal produce. For example, a Mediterranean-inspired dish featuring roasted vegetables, chickpeas, and a drizzle of olive oil highlights the benefits of whole ingredients, offering a meal that is both delicious and nutritious.

2. Incorporate Healthy Fats and Nutrient-Rich Oils

Healthy fats play a crucial role in supporting heart health, brain function, and satiety. Chefs can use nutrient-rich oils like olive oil, avocado oil, and flaxseed oil to add flavour and richness to dishes while providing essential fatty acids. In Mediterranean cooking, olive oil is a staple that enhances the taste of vegetables, salads, and fish. In Japanese cuisine, omega-3-rich fish like salmon and mackerel are celebrated for their health benefits. By incorporating these healthy fats, chefs can create dishes that are satisfying and aligned with health-focused eating.

3. Embrace Fermented Foods for Digestive Health

Fermented foods, such as yoghurt, kimchi, miso, and sauerkraut, are rich in probiotics that promote gut health. Chefs can incorporate fermented ingredients into their dishes to add complexity, tanginess, and nutritional benefits. For example, a chef might serve a side of kimchi with a grain bowl or add miso to a vegetable soup, providing diners with probiotics that support digestive wellness. Fermented foods also add depth of flavour, making them a valuable addition to health-conscious menus.

4. Use Spices and Herbs for Flavour and Health Benefits

Spices and herbs are rich in antioxidants and anti-inflammatory compounds, adding both flavour and health benefits to dishes. For instance, turmeric, widely used in Ayurvedic cooking, has potent anti-inflammatory properties, while ginger aids digestion and circulation. Chefs can experiment with spices like cumin, coriander, and cinnamon to create complex flavours that align with cultural traditions and promote wellness. Fresh herbs like basil, cilantro, and parsley also enhance the taste of dishes, providing essential vitamins and antioxidants.

5. Create Balanced Plates with Diverse Nutrients

Incorporating a variety of nutrient-dense ingredients into each dish ensures that diners receive a balanced meal. Chefs can follow the principles of "one soup, three sides" from washoku or create Mediterranean-inspired plates with lean proteins, vegetables, and whole grains. By including sources of protein, healthy fats, complex carbohydrates, and fibre, chefs can design dishes that support energy, satiety, and overall health.

6. Offer Customization Options for Dietary Preferences

Health-conscious diners often have specific dietary preferences, such as gluten-free, vegan, or low-carb options. Chefs can offer customization options to accommodate these preferences, allowing diners to choose dishes that align with their individual health goals. For example, a chef might offer a grain bowl with a choice of protein (tofu, chicken, or salmon) and a variety of vegetable toppings, providing a customizable, healthful meal. This flexibility enhances the dining experience, allowing each guest to enjoy a nutritious meal tailored to their needs.

7. Educate Diners on the Nutritional Benefits of Ingredients

Sharing information about the nutritional benefits of ingredients enhances the dining experience and builds trust with health-conscious diners. Chefs can include descriptions on the menu or have waitstaff share insights about the health benefits of specific dishes. For instance, a menu might highlight that a turmeric-infused dish supports inflammation reduction or that a dish containing fermented vegetables promotes gut health. Educating diners fosters an appreciation for health-focused dishes and encourages mindful, informed eating.

Creating nutritious, culturally inspired dishes allows chefs to cater to health-conscious diners while honouring culinary traditions. By using whole ingredients, embracing fermented foods, offering balanced plates, and educating diners, chefs can contribute to a dining experience that celebrates food as a source of health and wellness.

CHAPTER 30: CONCLUSION AND REFLECTIONS ON CULINARY ANTHROPOLOGY FOR MODERN CHEFS

Summary Of The Book's Themes And The Evolving Role Of Chefs In Cultural Preservation And Innovation

Throughout this book, we have traversed the landscapes of culinary anthropology, exploring how food functions as a bridge between cultures, a vehicle for tradition, and a tool for innovation. From ancient food traditions to modern gastronomic trends, each chapter has revealed that food is more than sustenance; it is a dynamic language that speaks of history, geography, spirituality, and identity. For chefs, understanding this language transforms cooking into a meaningful act—one that honours the legacy of culinary traditions while embracing the creative possibilities of innovation.

The role of chefs has evolved significantly over the centuries. Once seen primarily as skilled artisans working behind the scenes, today's chefs are cultural ambassadors, advocates for sustainability, and innovators who influence global food trends. Modern chefs have the opportunity to shape the way society perceives and experiences food, positioning themselves as stewards of both cultural preservation and progressive change. In this role, chefs are uniquely positioned to navigate the intersection of tradition and transformation, where the authenticity of inherited culinary practices meets the curiosity-driven spirit of experimentation.

A key theme that has emerged from our exploration is the dual responsibility of chefs to honour heritage and to innovate with respect. Culinary anthropology has shown us that food carries deep cultural significance and that each dish represents a lineage of knowledge, technique, and values. When chefs engage with global cuisines, they step into this lineage, taking on the role of storytellers who preserve and share the narratives embedded in each ingredient and recipe. At the same time, chefs are tasked with pushing boundaries, challenging conventions, and crafting new experiences that reflect contemporary tastes and concerns.

Another central theme is the ethical dimension

of cooking—the awareness that food choices have profound implications for people, animals, and the planet. As global food production faces unprecedented challenges, from environmental degradation to social inequalities, chefs are called to make decisions that prioritize sustainability, fairness, and respect for cultural heritage. This ethical responsibility is not just a matter of sourcing ingredients but of representing culinary traditions with integrity, honouring the diverse communities that have shaped these cuisines.

In many ways, the chef's journey mirrors the evolution of culinary anthropology itself. Just as culinary anthropology has moved from studying food as a static cultural artifact to viewing it as a living, adaptable expression of identity, chefs are reimagining their role from simple preparation to intentional representation. The modern chef stands at the crossroads of tradition and innovation, a custodian of culture, a creator of new flavours, and a mindful participant in the broader food ecosystem.

Reflections on the Ethical Responsibilities of Chefs in Representing Global Cuisines

As chefs engage with the diversity of global cuisines, they are entrusted with the responsibility of representing these traditions with honesty, respect, and humility. Cooking with cultural awareness means recognizing that each dish, ingredient, and technique carries a story, one that often reflects the struggles, triumphs, and values of the people who created it. The ethical responsibility of chefs lies in their ability to honour these stories, ensuring that cultural representation does not veer into appropriation or superficial imitation.

One of the primary ethical considerations in culinary representation is cultural authenticity. Authenticity does not necessarily mean rigid adherence to traditional recipes but rather an understanding of the context and significance behind each dish. For example, when preparing an Italian risotto, a Mexican mole, or a Japanese ramen, chefs should not simply replicate flavours but consider the traditions, skills, and regional nuances that make these dishes unique. This understanding creates a deeper connection between the chef and the cuisine, allowing them to interpret it with respect rather than reducing it to a trend.

However, as with any art form, there is room for reinterpretation and fusion in culinary practice. The challenge lies in balancing creativity with cultural sensitivity. Fusion cuisine can be a celebration of culinary diversity, but it must be approached with thoughtfulness and a commitment to preserving the integrity of each cultural element. When combining cuisines, chefs should ask themselves whether the fusion enhances the flavours and respect of each tradition or whether it risks reducing these elements to mere novelty. By framing innovation within the context of respect, chefs can create dishes that honour tradition while exploring new flavour possibilities.

Another ethical consideration is the social and environmental impact of ingredient sourcing. The choices chefs make in sourcing ingredients can either support sustainable practices or contribute to exploitation and degradation. Ethically sourced ingredients—whether fair trade coffee, locally farmed produce, or responsibly caught seafood—reflect a commitment to fairness, transparency, and sustainability. Chefs who prioritize ethical sourcing align themselves with a food system that respects both people and the planet, creating a kitchen that values integrity as much as flavour.

Chefs also play a role in addressing food justice by supporting local farmers, using ingredients that reflect seasonal availability, and educating diners on the importance of sustainable food choices. For instance, a chef who features indigenous grains or promotes plant-based dishes not only reduces their environmental impact but also raises awareness of food traditions that are often overlooked or marginalized. By incorporating these practices, chefs contribute to a food culture that values inclusivity, representation, and environmental stewardship.

Finally, chefs have the ethical responsibility to use their platform for positive change. Today's diners are more informed and socially aware than ever before, seeking dining experiences that align with their values. Chefs can educate their audience on issues such as food waste, ethical sourcing, and the cultural significance of each dish, transforming dining into an educational experience. By fostering awareness and empathy, chefs inspire diners to make mindful choices, promoting a more equitable and sustainable food system.

The ethical responsibilities of chefs extend beyond the kitchen, influencing the global food landscape and encouraging a culture of mindfulness, respect, and social responsibility. Through their choices, chefs have the power to shape a future where food is celebrated as both a personal experience and a shared cultural legacy.

Final Thoughts and Practical Advice for Chefs on Applying Culinary Anthropology Principles to Their Practice

As this journey through culinary anthropology concludes, chefs are left with the question: how can these principles of cultural appreciation, ethical responsibility, and innovation be applied in daily practice? Culinary anthropology offers chefs a roadmap for creating dishes that are not only flavourful but also meaningful—dishes that tell stories, build connections, and reflect a deep respect for the world's diverse food heritage. Below are practical ways for chefs to incorporate these principles into their culinary practice.

1. Approach Ingredients as Cultural Artifacts

Viewing ingredients as cultural artifacts shifts the way chefs interact with each element of a dish. Each spice, herb, or grain has a history, often rooted in specific climates, traditions, and cultural values. By researching and understanding the origin and significance of ingredients, chefs can deepen their appreciation and create dishes that resonate with authenticity. For example, using turmeric in a dish is not just about flavour; it is about honouring its role in Ayurvedic healing and its symbolic significance in Indian culture. This mindful approach transforms cooking into a respectful dialogue with the past and the cultures that have contributed to global cuisine.

2. Experiment with Integrity and Cultural Sensitivity

Innovation is at the heart of culinary artistry, but it must be balanced with cultural sensitivity. When experimenting with new flavours or creating fusion dishes, chefs should consider whether their choices enhance or dilute the essence of each cuisine. By focusing on techniques that complement rather than overshadow, chefs can create fusion dishes that feel cohesive and respectful. Consulting with cultural experts, studying traditional methods, and incorporating elements thoughtfully are essential steps for creating fusion that celebrates rather than appropriates.

3. Design Menus That Reflect Ethical and Sustainable Choices

Menus are powerful tools for showcasing a chef's values. By designing menus that prioritize local, seasonal, and ethically sourced ingredients, chefs demonstrate their commitment to sustainability and responsible sourcing. Seasonal menus celebrate nature's cycles, offering diners a taste of what is freshest and most sustainable. Ethical sourcing also extends to the choice of proteins and produce, favouring suppliers who practice fair labour and environmental stewardship. By aligning their menus with these principles, chefs contribute to a food culture that respects both the environment and the communities that sustain it.

4. Use Storytelling to Enhance the Dining Experience

Storytelling is a powerful way to share the cultural significance of each dish with diners. Chefs can use menus, conversations, or digital platforms to explain the origin of ingredients, the inspiration behind recipes, or the cultural traditions that influence their cooking. For example, a menu might highlight that a dish was inspired by a specific region, using spices, cooking methods, or presentation styles traditional to that area. By incorporating storytelling, chefs create a richer, more immersive experience, transforming dining into a journey through history, culture, and flavour.

5. Embrace Continuous Learning and Cultural Curiosity

Culinary anthropology is a field that encourages lifelong learning. Chefs can deepen their knowledge by studying global cuisines, attending workshops, or collaborating with chefs from different cultural backgrounds. This commitment to learning fosters curiosity and respect for diverse food traditions, allowing chefs to evolve and expand their culinary repertoire. Embracing cultural curiosity not only enhances a chef's skillset but also strengthens their role as a cultural ambassador, bridging culinary traditions with modern interpretations.

6. Advocate for Inclusivity and Representation in the Culinary World

Finally, chefs have the unique opportunity to advocate for inclusivity and representation within the culinary industry. By supporting underrepresented cuisines, ingredients, and voices, chefs contribute to a more inclusive food culture. This can mean featuring dishes inspired by Indigenous, regional, or lesser-known cuisines or collaborating with chefs who specialize in these areas. Inclusivity also extends to the kitchen staff, creating a work environment that values diverse perspectives and backgrounds. Chefs who champion inclusivity make a lasting impact, helping to build a culinary world that celebrates all cultures.

Culinary anthropology equips chefs with the tools to cook with intention, respect, and creativity. By embracing the principles of cultural awareness, ethical responsibility, and innovation, chefs can transcend the role of cook to become cultural custodians, storytellers, and advocates for a sustainable future. The journey through culinary anthropology is one of continuous growth and reflection, where each dish becomes an opportunity to honour the past, celebrate the present, and envision a better future for food.

As chefs step forward, they carry with them the wisdom of ancient traditions, the potential of modern innovation, and the responsibility to create a culinary legacy that respects the planet and its people. This is the power of culinary anthropology—a discipline that transforms food into a meaningful, universal experience, connecting chefs and diners in a shared appreciation of the world's rich and diverse culinary heritage.

EPILOGUE: A JOURNEY BEYOND THE PLATE

As we come to the close of this exploration, it is clear that the journey through culinary anthropology is one without end—a perpetual invitation to discover, to innovate, and to deepen one's connection to the world through food. This book, with its chapters on tradition, innovation, ethics, and sustainability, has merely opened the door to a vast, interconnected world of culinary meaning. For chefs, culinary anthropology is a roadmap, a reminder that every meal, every dish, and every ingredient is a part of something greater: a history, a culture, a collective story of humanity.

Food as Legacy

The food we create today is not just for the present; it is a legacy for the future. Just as generations before us have passed down recipes, techniques, and culinary wisdom, today's chefs are building their own contributions to this tapestry of taste. They are caretakers of ancient knowledge, tasked with preserving it while allowing it to evolve to meet the needs of a changing world. When a chef prepares a dish inspired by an ancient practice or distant land, they become a link in this ongoing chain, respecting the legacy that shaped them and crafting something that will, in turn, shape future generations.

Food's power lies in its ability to transcend boundaries. It can carry memories, evoke emotions, and communicate values that words alone cannot express. When chefs embrace this perspective, they elevate their craft beyond technique and flavour; they engage in a form of art that touches lives, builds connections, and enriches our shared human experience. The culinary world is expansive, and it is constantly transforming. Yet, amid the evolution, the importance of legacy remains. It is this legacy that each chef contributes to with every dish they create, with each choice they make in the kitchen.

The Power of Curiosity and Respect

As chefs continue their journey, one of the most valuable companions they can bring along is curiosity. Curiosity drives discovery; it leads chefs to explore ingredients they have never used, to study cooking traditions unfamiliar to them, and to question the potential of each flavour they encounter. With curiosity comes respect—a deep appreciation for the people, places, and traditions behind each dish. When these two elements merge, curiosity and respect transform culinary creation into a meaningful act of cultural celebration.

The richness of the culinary world lies in its diversity, and true respect means recognizing the weight and beauty of that diversity. It means taking the time to understand the ingredients that others may hold sacred, the recipes that have endured for centuries, and the customs that are woven into the fabric of other societies. This respect cultivates an awareness that every dish, whether simple or complex, is part of a story far greater than the chef alone. Through curiosity and respect, chefs become both students and teachers of the world's culinary heritage, sharing the beauty of global food cultures while honouring the people who have sustained them.

Innovation as a Path to Preservation

Throughout this book, we have explored how tradition and innovation can coexist. Innovation is not a departure from heritage; it is, in many ways, a way to preserve and adapt it for future generations. When chefs innovate, they are not abandoning tradition but carrying it forward, infusing it with new perspectives and techniques that speak to the world of today. Innovation can honour tradition by expanding its reach, making ancient flavours relevant and accessible to a new audience. It is through innovation that the past continues to breathe, evolving without losing its core essence.

However, this journey demands balance. Chefs must innovate with integrity, ensuring that their creations pay homage to the essence of the original, even as they interpret it through a modern lens. This path of responsible innovation requires both creativity and humility—a recognition that while every chef has a unique voice, they are also part of a larger narrative. In the fusion of past and present, chefs find the opportunity to create dishes that are not only new but also resonant with history, allowing diners to experience both freshness and familiarity.

A Community of Storytellers

Every chef, in their own way, is a storyteller. Through food, they tell stories of origin, migration, adaptation, and memory. A dish can tell of a family's journey across continents, the ingenuity of a community that turned local ingredients into delicacies, or the wisdom of a culture that saw food as medicine. Chefs are, thus, not merely preparing food; they are sharing the lives, struggles, and triumphs of people from all corners of the earth.

But these stories are not told in isolation; they are part of a community narrative, where every chef's story contributes to a broader understanding of humanity. The act of cooking connects chefs to each other and to the world. Whether in bustling cities or quiet rural towns, every kitchen becomes a place where stories are created, shared, and celebrated. It is a reminder that chefs, despite their individual creativity, are united by a shared mission: to bring people together, to foster understanding, and to celebrate the beauty of diversity through food.

Carrying Culinary Anthropology Forward

In closing, culinary anthropology is not merely an academic study; it is a way of seeing the world. It teaches chefs to be mindful of the interconnectedness of cultures, the ethical responsibilities that come with culinary influence, and the potential for food to inspire and unite. As chefs carry these lessons forward, they will discover new ways to honour the world's culinary heritage, from sourcing ingredients with care to crafting dishes that invite diners to connect with other cultures.

As the culinary world continues to evolve, there will be challenges and uncertainties. New trends will emerge, ethical dilemmas will arise, and chefs will need to adapt to meet the needs of an ever-changing world. But if they carry with them the principles of culinary anthropology—curiosity, respect, integrity, and a commitment to cultural representation—they will always find themselves anchored, able to navigate the future with purpose.

So, to all the chefs embarking on this journey, remember that you are more than creators of flavour. You are curators of tradition, custodians of culture, and champions of connection. You have the power to make every meal a celebration of diversity, every dish a lesson in history, and every bite an invitation to explore. Let the lessons of culinary anthropology guide your steps, enriching your craft and connecting you ever more deeply to the world and to the people who sustain it.

Here's to a world where food is a bridge, chefs are storytellers, and every meal is a shared journey across cultures, generations, and hearts.

REFERENCES

Anderson, E. N. (2005). Everyone Eats: Understanding Food and Culture. New York University Press.

Belasco, W., & Scranton, P. (Eds.). (2002). Food Nations: Selling Taste in Consumer Societies. Routledge.

Bestor, T. C. (2004). Tsukiji: The Fish Market at the Centre of the World. University of California Press.

Chang, H.-J. (2019). Edible Economics: How to Eat Your Way Through Global Food History. Bloomsbury.

Counihan, C., & Van Esterik, P. (Eds.). (2013). Food and Culture: A Reader (3rd ed.). Routledge.

Curtin, D. W., & Heldke, L. M. (Eds.). (1992). Cooking, Eating, Thinking: Transformative Philosophies of Food. Indiana University Press.

Douglas, M. (1972). Deciphering a Meal. Daedalus, 101(1), 61-81.

Flandrin, J.-L., & Montanari, M. (Eds.). (1999). Food: A Culinary History from Antiquity to the Present. Columbia University Press.

Goldstein, D., & Merkle, K. (Eds.). (2015). Culinary Tourism. University Press of Kentucky.

Heldke, L. M. (2003). Exotic Appetites: Ruminations of a

Food Adventurer. Routledge.

Kiple, K. F., & Ornelas, K. C. (Eds.). (2000). The Cambridge World History of Food. Cambridge University Press.

Krondl, M. (2008). The Taste of Conquest: The Rise and Fall of the Three Great Cities of Spice. Ballantine Books.

Mintz, S. W. (1985). Sweetness and Power: The Place of Sugar in Modern History. Viking.

Mouritsen, O. G., & Styrbæk, K. (2014). Umami: Unlocking the Secrets of the Fifth Taste. Columbia University Press.

Narayan, K. (1995). Eating Cultures: Incorporation, Identity, and Indian Food. Social Text, 43, 35-50.

Nestle, M. (2013). Food Politics: How the Food Industry Influences Nutrition and Health. University of California Press.

Pollan, M. (2006). The Omnivore's Dilemma: A Natural History of Four Meals. Penguin Press.

Scapp, R., & Seitz, B. (Eds.). (1998). Eating Culture. State University of New York Press.

Schlosser, E. (2002). Fast Food Nation: The Dark Side of the All-American Meal. Mariner Books.

Shurtleff, W., & Aoyagi, A. (2012). History of Miso, Soy Sauce, and Tamari. Soyinfo Centre.

Sutton, D. E. (2001). Remembrance of Repasts: An Anthropology of Food and Memory. Berg.

Wilson, C. A., & Rathje, W. L. (Eds.). (2012). The

Archaeology of Food and Cuisine. University Press of Florida.

Wulf, A. (2015). The Invention of Nature: Alexander von Humboldt's New World. Alfred A. Knopf.

Additional References on Sustainability and Culinary Ethics

Gunders, D. (2012). Wasted: How America Is Losing Up to 40 Percent of Its Food from Farm to Fork to Landfill. Natural Resources Defense Council.

Parasecoli, F., & Johnston, J. (Eds.). (2017). Feeding the City: Work and Food Culture of the Urban Poor. Bloomsbury Academic.

Rockefeller, A., & Burke, M. (2016). Permaculture for a New World. Chelsea Green Publishing.

Rose, R., & Labonte, M. (2009). Indigenous Food Sovereignty in Canada. International Journal of Indigenous Health, 5(1), 5–13.

Stein, K. (2017). Gastronomy and the Kitchen of Tomorrow: Ethical and Sustainable Perspectives on Future Food. University of Gastronomic Sciences Press.

Symons, M. (1994). A History of Cooks and Cooking. University of Illinois Press.

GLOSSARY OF TERMS

Agriculture
The science, art, and practice of cultivating soil, growing crops, and raising animals for food, fibre, and other products.

Ahimsa
A principle in Hinduism, Buddhism, and Jainism that emphasizes non-violence and respect for all living beings. Ahimsa influences dietary practices, particularly vegetarianism, in Indian culture.

Authenticity
In the culinary context, the concept of preparing food that remains true to the original methods, ingredients, and cultural context of a dish or cuisine.

Ayurveda
An ancient Indian system of medicine that emphasizes balance among bodily systems and uses diet, herbal treatments, and yogic breathing to promote health.

Biodiversity
The variety of plant and animal life in the world or a particular ecosystem. Biodiversity in food systems promotes resilience, health, and sustainable

agriculture.

Chinampas
A traditional Mesoamerican agricultural technique, often called "floating gardens," where crops are grown on small, rectangular plots in shallow lake beds.

Culinary Anthropology
The study of food in its cultural context, exploring how food reflects social norms, identity, history, and the relationship between food practices and human behaviour.

Cultural Appropriation
The adoption or use of elements of one culture by members of another culture in a way that can be disrespectful, misrepresentative, or commodifying.

Cultural Relativism
The principle of understanding and respecting the practices, beliefs, and values of other cultures without judging them by the standards of one's own culture.

Dosha
In Ayurveda, the three bodily energies (Vata, Pitta, and Kapha) believed to govern physiological and psychological processes, influencing dietary needs and lifestyle practices.

Ethical Sourcing
The practice of procuring ingredients that are produced in a way that respects social, environmental, and economic factors, ensuring fairness and sustainability in food production.

Farm-to-Table

A movement that emphasizes sourcing food directly from local farms, promoting seasonal and sustainable ingredients in restaurants and reducing the distance food travels to the consumer.

Fermentation
A food preservation process where microorganisms like yeast and bacteria convert sugars into acids, gases, or alcohol. Fermented foods (e.g., kimchi, sauerkraut, miso) are valued for their probiotic benefits.

Fusion Cuisine
A style of cooking that combines elements from different culinary traditions to create new dishes, often blending ingredients, techniques, and flavours from multiple cultures.

Gastronomy
The art and science of good eating, including the study of food, culture, and culinary traditions, as well as the exploration of new flavours and cooking methods.

Globalization
The increasing interconnectedness of the world's economies, cultures, and populations, facilitated by trade, technology, and cultural exchange. In culinary terms, it refers to the spread of cuisines, ingredients, and food practices across borders.

Ichiju Sansai
A Japanese dining principle meaning "one soup, three sides," reflecting a balanced meal structure that emphasizes variety, seasonal ingredients, and portion control.

Indigenous Food Sovereignty

The right of Indigenous communities to control their food systems, prioritize traditional practices, and access culturally appropriate, sustainably harvested foods.

Lab-Grown Meat

Also known as cultured or cell-based meat, this is meat produced by cultivating animal cells in a lab environment, aiming to reduce environmental impact and animal suffering.

Mediterranean Diet

A dietary pattern originating from the Mediterranean region, focusing on vegetables, fruits, whole grains, fish, olive oil, and limited meat, associated with health benefits such as heart health and longevity.

Molecular Gastronomy

A subdiscipline of food science that explores the physical and chemical transformations of ingredients during cooking, often used to create innovative textures and flavours.

Nikkei Cuisine

A fusion cuisine that blends Japanese and Peruvian culinary traditions, born from Japanese immigrants adapting their cuisine to local Peruvian ingredients.

Permaculture

A system of agricultural and social design principles centred around simulating or directly utilizing the patterns and features observed in natural ecosystems, promoting sustainable food production.

Probiotics

Live bacteria and yeasts that are beneficial for digestive

health, commonly found in fermented foods like yoghurt, kimchi, and sauerkraut.

Sustainable Agriculture
Farming practices that maintain the productivity of land while preserving environmental health, often emphasizing reduced chemical use, soil conservation, and water management.

Symposium
In ancient Greece, a social gathering where individuals discussed philosophy, politics, and culture over food and wine, illustrating the cultural role of food as a means of intellectual and social exchange.

Terroir
A French term describing the unique characteristics imparted to food or wine by the geography, soil, climate, and local environment of its origin.

Three Sisters
A traditional Native American planting method that interplants corn, beans, and squash. Each plant supports the growth of the others, reflecting Indigenous knowledge of sustainable agriculture.

Traditional Chinese Medicine (TCM)
An ancient Chinese medical system that uses herbal remedies, diet, and practices like acupuncture to balance the body's Qi (vital energy) and maintain health, often categorizing foods by their warming or cooling properties.

Umami
Known as the "fifth taste," umami is a savoury flavour found in foods like mushrooms, soy sauce, tomatoes,

and aged cheeses. It is an essential component in many culinary traditions and enhances flavour complexity.

Zero-Waste Cooking
A culinary approach that aims to minimize food waste by utilizing every part of an ingredient, finding creative ways to repurpose leftovers and byproducts, and promoting sustainable resource use.

ACKNOWLEDGEMENTS

This book, _Culinary Anthropology: Gastronomy and Culture for Chefs_, is the culmination of many years of fascination with the intricate relationship between food, culture, and history. I am deeply grateful to all those who have inspired and supported this journey.

First and foremost, I would like to thank the chefs, culinary experts, and anthropologists whose work has been instrumental in shaping my understanding of food and culture. Their insights, dedication to culinary craft, and profound respect for food's cultural significance have been a guiding light throughout this project.

I extend my heartfelt gratitude to my family and friends, whose patience, encouragement, and unwavering belief in my work allowed me to bring this book to life. Their support has been invaluable, particularly during the long hours of research, writing, and revisions.

I am also grateful to my publisher, Irene Minds, for recognizing the importance of this book and helping it reach readers who are passionate about culinary heritage and gastronomy. To the editorial and design teams, thank you for your professionalism and expertise in transforming this manuscript into a polished final work.

Finally, to all the readers and aspiring chefs who seek to understand food beyond taste and technique—this book is for you. May it inspire you to explore, honour, and celebrate the cultural legacy behind every dish.

Thank you all for being part of this journey.

Dr Bhaskar Bora
November 2024

COPYRIGHT INFORMATION

Culinary Anthropology: Gastronomy and Culture for Chefs
© 2024 Dr Bhaskar Bora
All rights reserved.

Published by Irene Minds

First Edition: 2024

Publisher: Irene Minds

DISCLAIMER

This book is intended for educational and informational purposes only. The author and publisher have made every effort to ensure the accuracy and completeness of the information provided in this book; however, they assume no responsibility for errors, inaccuracies, or omissions. The recommendations, practices, and views expressed in this book are based on the author's research and perspectives and do not constitute professional advice.

Readers should use this book as a resource to gain insight into culinary anthropology but should seek additional resources, professional guidance, or expert consultation if applying any of the methods, recommendations, or techniques described herein. Neither the author nor the publisher is liable for any outcomes, direct or indirect, arising from the use or application of the contents of this book.

The author and publisher are not responsible for any third-party websites or resources mentioned or linked in this book. References to brands, companies, or organizations are for informational purposes only and do not imply endorsement by the author or publisher.